AF699232

AN INTRODUCTION TO PARALLEL COMPUTATIONAL FLUID DYNAMICS

AN INTRODUCTION TO PARALLEL COMPUTATIONAL FLUID DYNAMICS

Sauro Succi

and

Francesco Papetti

Nova Science Pubishers, Inc.

New York

Art Director: Maria Ester Hawrys
Assistant Director: Elenor Kallberg
Graphics: Susan A. Boriotti, Frank Grucci and Kerri Pfister
Manuscript Coordinator: Phylis Gaynor
Book Production: Gavin Aghamore, Joanne Bennette, Michele Keller, Christine Mathosian and Tammy Sauter
Circulation: Iyatunde Abdullah, Cathy DeGregory and Annette Hellinger

Library of Congress Cataloging–in–Publication Data
available upon request

ISBN 1-56072-354-8

© 1996 Nova Science Publishers, Inc.
6080 Jericho Turnpike, Suite 207
Commack, New York 11725
Tele. 516-499-3103 Fax 516-499-3146
E Mail Novasci1@aol.com

All rights reserved. No part of this book may be reproduced, stored in a retrieval system or transmitted in any form or by any means: electronic, electrostatic, magnetic, tape, mechanical, photocopying, recording or otherwise without permission from the publishers.

Printed in the United States of America

To our wives

Claudia and Alessandra

Contents

Preface vii

1 Fluid dynamics research in science and engineering 1

1.1 Introduction 1
1.2 Turbulence 2
1.3 The role of geometry 7
1.4 Reactive Flows 9
1.5 The role of computers 10

2 The basic equations of fluid flow 15

2.1 The Navier-Stokes equations 18
2.2 Various forms of the Navier-Stokes-Fourier equations 21
2.2.1 Eulerian form 21
2.2.2 The Lagrangian formulation 23
2.2.3 The vorticity-streamfunction formulation 26
2.2.4 Potential flows 29

3 Numerical solution of fluid flow equations: Grid Methods 31

3.1 Grid methods vs. Particle methods 32
3.2 Grid methods 34
3.2.1 Finite difference methods 35
3.2.2 The Finite Volume method 38
3.2.3 The Finite Element method (FEM) 43
3.2.4 Spectral methods 50
3.3 Time-discretization and non-linearity 53
3.3.1 Non-linearity 57
3.4 Solving linear algebraic equations 57
3.4.1 Direct methods 58

3.4.2 Iterative methods 59

4 Numerical solution of fluid flow equations: particle methods 65

4.1 Particle methods 65

4.1.1 Molecular dynamics 66

4.1.2 Direct Monte Carlo Simulation 69

4.1.3 Lattice gas 72

4.1.4 Lattice Boltzmann 76

4.1.5 Vortex methods 78

4.2 Smoothed Particle Hydrodynamics 80

4.3 Time marching with particle methods 82

4.3.1 Molecular Dynamics 82

4.3.2 Lattice Gas and Lattice Boltzmann 82

4.3.3 Direct Simulation Montecarlo 83

4.3.4 Vorton methods 83

4.3.5 Smoothed-Particle Hydrodynamics 84

5 Parallel computing 85

5.1 Vector computers 88

5.2 Superscalar computers 92

5.3 Parallel architectures 93

5.3.1 Control mechanism 94

5.3.2 Memory organization 95

5.4 Topology 97

5.5 Parallel programming: general concepts 102

5.5.1 Scalability 103

5.5.2 Fraction of parallel content 105

5.5.3 Load balance 108

5.5.4 Coordination and communication overheads 109

5.6 Redundancy 116

5.7 Summary 116

6 Parallel Explicit Methods 119

6.1 Parallel Explicit Grid Techniques 121

6.2 Structured Explicit (SE) 121

6.2.1 Distributed memory MIMD 125

6.2.2 Distributed memory SIMD 128

6.2.3 Efficiency of structured explicit schemes 128

6.3 Unstructured explicit (UE) 130

6.3.1 Parallelization of Matrix-Vector-Multiply 138
6.3.2 Efficiency of Unstructured Explicit schemes 141
6.4 Parallel Spectral Methods 141
6.5 Particle Methods 147
6.5.1 Parallel Molecular Dynamics 148
6.5.2 Parallel Direct Simulation Monte Carlo 155
6.5.3 Lattice Gas 155
6.5.4 Lattice Boltzmann 157
6.5.5 Vortex Methods (VM) 158

7 Parallel Implicit Methods **165**
7.1 Parallel iterative schemes 165
7.1.1 Relaxation schemes 165
7.1.2 Multicolored Block-Iterative Methods 172
7.2 Gradient Methods 175
7.2.1 LU Decomposition 179
7.2.2 Parallelizing the backward/forward substitution 182
7.2.3 Combining direct and iterative methods 184
7.2.4 Direct solvers for unstructured implicit schemes 188
7.2.5 Incomplete nested dissection 190
7.3 A pause of reflection 192
7.3.1 Explicit / Implicit 193
7.3.2 Iterative/Direct 193
7.3.3 Structured/Unstructured 193
7.4 Spectral Methods 194
7.5 Particle Methods 194

8 Case studies in parallel CFD **197**
8.1 Lattice Boltzmann on the IBM SP1 197
8.2 Engine combustion with unstructured conjugate gradient solver 202
8.3 Parallel Spectral Codes for Homogeneous Incompressible Turbulence on the CM-2 205
8.4 Incompressible flow past solid bodies with Parallel Vortex Methods 207
8.5 Parallel Rarefied Gas Dynamics 209

A Parallel computer architectures **215**
A.1 The IBM SP1/SP2 architecture 215
A.2 The nCUBE architecture 216

A.3 The Intel Paragon XP 217
A.4 The CM-5 architecture 218
A.5 The Cray T-3D architecture 218
A.6 The SGI Power Challenge architecture 219

Bibliography **220**

Subject Index **233**

Preface

This book provides an introduction to the subject of Parallel Computational Fluid Dynamics (CFD), namely the scientific discipline aimed at solving fluid dynamics problems using parallel computers. Parallel CFD is a young and fast growing subject, whose saturation is not yet in sight.

And yet, fueled by the impressive advances of computer technology on the one side, and the unceasing progress of numerical techniques on the other, Parallel CFD is rapidly getting to the point where it can be viewed as a cost-effective alternative and/or complement to direct experimentation in the study of many fluid-dynamics applications of academic and industrial interest.

In view of the above, it appears that there's scope for making a wide audience of potential future users acquainted with the basic ideas behind Parallel CFD.

Basic ideas, rather than in-depth details, are indeed the main focus of this work, which is consequently targeted to CFD practicioners with no specific prior knowledge of high-performance computing in general, and parallel computing in particular. It is hoped that interests triggered by this book may stimulate further reading of specialized topics of specific relevance to each reader.

Acknowledgements

The material collected in this book is the outcome of the experience gained over many years of fruitful interaction with our colleagues at the IBM European Centre for Scientific and Engineering Computing and with many other academic and industrial partners all over Europe.

They have all been our teachers.

We feel especially indebted to M.Bernaschi, M.Briscolini, P.Santangelo and M.Vitaletti for several enlightening discussions. Special thanks are due to Proff. S.Orszag, R.Piva and R.Hockney for the many valuable suggestions. Finally we wish to acknowledge G.Bachler and AVL List in Graz for granting us with permission to reproduce Fig. 8.5, 8.6.

Chapter 1

Fluid dynamics research in science and engineering

1.1 Introduction

Fluid-dynamics is the scientific discipline dealing with the quantitative description of the behaviour of the fluid state of matter, i.e. gases and liquids.

Fluids play a crucial role in many aspects of our daily life: the air we breath is a fluid, and so is the blood flowing in our veins, just to mention two but a few examples of immediate interest. No less important is the role of fluids on a more global scale; think of the dynamics of the atmosphere or to water motion in the Oceans, two phenomena of utmost importance for the life of our planet. Fluid behaviour is also encountered down to the microscopic scales; under specific conditions the electrons in a wire of copper behave like a conducting fluid carrying electric current.

In view of their pervasive presence across virtually all of human activities, there is little doubt that the fluid state of matter is definitely worth being studied in full depth. In this context, it comes to no surprise that the basic equations governing fluid dynamics have been laid down more than a century ago by the french engineer Louis Marie Henri Navier and the irish physicist sir George Gabriel Stokes [1].

Although the Navier-Stokes equations have been known for more than a century, many of the important features of fluid flows continue to elude our thorough comprehension.

In a word, under many conditions of practical interest, fluid behaviour is still **unpredictable**.

Weather forecasting is probably the most immediate application which typifies this unpredictability [2].

Why is this so ? The point is that fluids have been given by Nature a tremendous amount of freedom, and even exploiting a small share of this freedom is more than enough to generate a wide variety of extremely complex behaviours. In a more technical language, this ability to exhibit complex behaviour is referred to as to **turbulence**. Quoting Karniadakis and Orzsag [3], "Understanding turbulent flows is a *grand challenge* comparable to other prominent scientific problems such as the large-scale structure of the Universe or the nature of subatomic particles". Let us see why.

1.2 Turbulence

Turbulence is not an intrinsic property of fluids, but rather a specific (and actually the most common) state of fluid **flows**. This state is entered whenever a sufficiently high number of degrees of freedom take an active part in the dynamics of the fluid flow.

Having many degrees of freedom in action, alone, is by no means a sufficient condition to produce turbulent behaviour. The most salient feature is that these degrees of freedom must be interacting in a **non-linear** fashion.

The creative power of non-linear systems has been recognized since the early 60's, with the seminal work of Lorenz, Ruelle and other pioneers of non-linear science [4].

Possibly, the most distinctive feature of non-linear dynamical systems is that their physics is coupled at all scales of motion. This is in a blatant contrast with the much more comfortable behaviour of linear systems, in which each scale of motion evolves on its own, in a complete isolation with respect to other scales.

This inter-scale coupling lies at the very heart of fluid turbulence. That's why, according to Frisch and Orszag [5]

"At the present time, we know less about the fine-scale of turbulence (at, for example, 1 *mm* in the atmosphere) than about the structure of atomic nuclei."

The degree of turbulence of a given flow is usually measured by a single dimensionless parameter, the Reynolds number, defined as

$$Re = \frac{UL}{\nu} \tag{1.1}$$

where U is a typical flow speed, L a typical macroscopic flow scale, and ν is the molecular viscosity of the fluid.

Based on dimensional considerations on the Navier-Stokes equations, the number of degrees of freedom active in a turbulent flow at Reynolds Re is given by

$$N_{dof} \simeq Re^{9/4} \tag{1.2}$$

Given the fact that many fluids have $\nu \simeq 10^{-2} cm^2/s$, and many flows of practical interest work at $U > 1\, m/s$ and $L > 1\, m$, typical Reynolds numbers in real life are well in excess of 10^6, with a corresponding number of degrees of freedom of the order of 10^{13} !

Why are most flows turbulent, i.e. why is the Reynolds number generally so large? To answer this question, it is useful to focus on the physical interpretation of the Reynolds number.

In physical terms, Re is a measure of the relative strength of **convection** (UL) versus **dissipation** (ν). Now, while convection occurs at a macroscopic scale, dissipation becomes active only at the **molecular** level. This interpretation neatly emerges by rewriting the equation 1.1 in terms of microscopic quantities, like the molecular mean speed, v, and the molecular mean free path l_{mfp}, i.e. the distance walked by a molecule between two collisions with other molecules.

From ordinary kinetic theory of gases, we know that the fluid viscosity is given by [6]

$$\nu \simeq v l_{mfp} \tag{1.3}$$

in such a way that the equation 1.1 becomes

$$Re \simeq \frac{U}{v} \frac{L}{l_{mfp}} \tag{1.4}$$

Since the molecular mean speed is of the order of the sound speed, c_s, the equation 1.4 can also be written as

$$Re \simeq \frac{Ma}{Kn} \tag{1.5}$$

where $Ma = U/c_s$ is the flow Mach number and $Kn = l_{mfp}/L$ is the Knudsen number, c_s denoting the speed of sound.

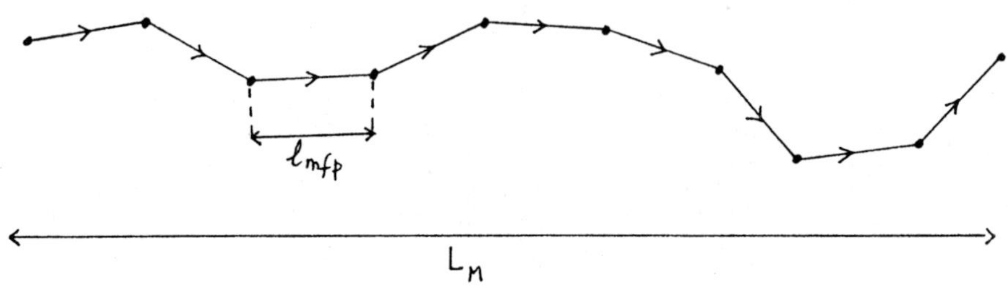

Figure 1.1: A typical macroscopic variation length (L_M) uncovers several mean free-pathes (l_{mfp})

For many fluid applications $Ma \simeq O(1)$, while $Kn \simeq 10^{-6}$. The conclusion is that the Reynolds number is high because it reflects the number of mean-free paths contained in the macroscopic length L.

In a pictorial sense, Re measures the length of the bridge connecting the **macroscopic** to the **microscopic** worlds. The way how this bridge can "stand" in a stable fashion is deeply rooted in the peculiar features of turbulence energetics. A qualitative picture of flow energetics looks as follows: The energy fed at large scales (typically via boundary conditions) is transferred to increasingly smaller scales via the non-linear interaction between the degrees of freedom active in the flow, in a sort of "cascade" from large to small eddies.

This energy cascade stops once the smallest scales become comparable in size with the typical scale below which dissipation takes over and the energy carried on by organized motion is irreversibly dissipated into heat (Fig. 1.2).

The shortest scale active in the flow is universally known as "Kolmogorov scale", η, after the russian matematician Andrey Kolmogorov who laid the foundations of turbulence back in the early 40's [7].

According to Kolmogorov theory, the dissipation scale η is related to the

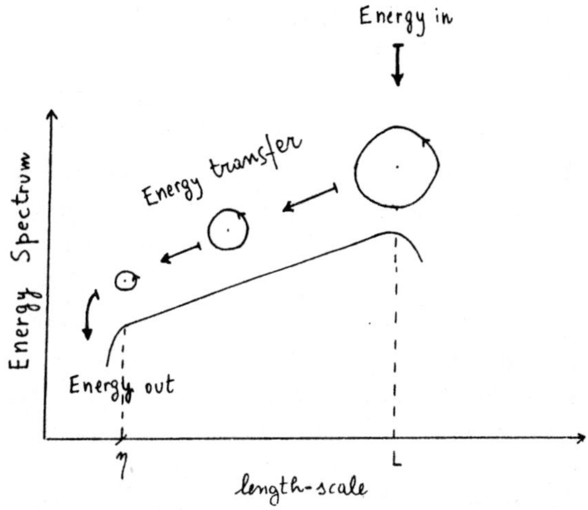

Figure 1.2: Spectrum of the energy distribution across the various scales of motion in a turbulent flow

macroscale L by the following relationship:

$$\eta \simeq \frac{L}{Re^{3/4}} \tag{1.6}$$

This equation shows that the "energy bridge" can span several decades. Take for example $Re \simeq 10^8$, a common value in atmospheric turbulence, the energy fed at, say, $1\,km$ scales can still be found at scales as small as $1\,mm$!

The way how a turbulent flow is able to keep the energy "frozen" across such a wide region of scales is one of the outstanding questions of turbulence research.

Although no single theory has emerged yet as the clear winner, there is a widespread consensus in the scientific community on the central role played by **coherent structures**.

Coherent structures are regions of fluid flow in which turbulence reorganizes itself into regular, non-chaotic motion. Typically, they take the form of spiralizing structures, called vortices, characterized by a rapid rotation of the flow (think of a tornado as of a typical large-scale coherent structure).

The distinctive feature of coherent structures is that they do not dissipate energy; as a result, they serve as sort of (meta)stable energy storage devices which allow energy to be passed across different scales without being affected by dissipation.

Like in many other non-linear systems, coherent structures seem to be organized in hierarchical fashion, with large vortices breaking down into a number of smaller "child" vortices, which, in turn, break down into still smaller vortices and so on, down to the Kolmogorov length.

Another leading avenue in theoretical fluid dynamics research is the study of the effects of small-scales on the large-scale dynamics of the flow. This subject is of utmost importance, for the study of transport phenomena within fluid flows. It should be appreciated that, on the practical side, the most immediate effect of fluid turbulence is **enhanced transport**: diffusion of pollutants in the atmosphere and cream stirring in your cup of coffee are two -disparate- but equally representative examples.

Here, again, coherent-structures play a leading role. While it is true that these eddies conserve energy, it's equally true that as they move around the flow, they can transfer their energy between different regions of the flow much more effectively than molecular motions: whence the denomination of enhanced transport.

A guiding idea, due to Prandtl, is that a turbulent flow can be paralleled to a "gas of eddies", the smallest eddies acting on the large ones in a diffusive manner. This picture, clearly inspired to kinetic theory of gases, leads to the introduction of the important concept of **eddy viscosity** ν_{eddy} [8].

Denoting by v_{rms} the rms fluctuating velocity and l_M a typical correlation scale ($l_M \simeq L$ since coherent structures cover the macro-scales of the flow) the eddy-viscosity can be estimated of the order of

$$\nu_{eddy} \simeq v_{rms} L \tag{1.7}$$

Once compared with equation 1.1, equation 1.7 yields ($v_{rms} \simeq U$)

$$\frac{\nu_{eddy}}{\nu} \simeq Re \tag{1.8}$$

This relation shows that the effective transport rates in a turbulent flow are enhanced by approximately a factor Re with respect to sheer molecular motion.

The eddy-viscosity concept proves very useful in the description of many practical flows.

This is somewhat surprising because most of those flows are far-out of equilibrium, and, as a result, they badly violate the "quasi-equilibrium" assumption which underlies the very concept of eddy viscosity.

Eddy-viscosity is possibly the simplest form of turbulence **closure**. We already pointed out that a better comprehension of the effect of small-scale dynamics on large-scale structures is one of the outstanding problems of fluid-dynamics.

Unfortunately there is no way to solve this problem in an exact, definitive and closed form. The point is that each spatial scale is significantly affected by a smaller one, which in turn is affected by an even smaller one, and so on down the line in a sort of an endless loop known as the "closure problem".

To any practical purpose, it is clear that at some point this endless and self-repeating hierarchy has to be broken by means of some "ad-hoc" assumption on the nature of the flow.

For instance, in the case of Prandtl eddy-viscosity, the assumption is that the average fluctuating momentum flux tensor $\tau_{ij} =< u_i' u_j' >$ (Reynolds stress tensor) is directly proportional to the large-scale stress tensor $S_{ij} = (\frac{\partial U_j}{\partial x_i} + \frac{\partial U_i}{\partial x_j})$ via a scalar function of S_{ij} which is precisely the eddy-viscosity in point: $\tau_{ij} = \nu(S) S_{ij}$.

The rationale behind this assumption is that turbulent fluctuations should be particularly high and active there where the large-scale flow exhibits strong gradients (i.e. source of instability).

As we already remarked, this assumption sounds pretty plausible, and in fact it proves quite useful in a number of practical situations. Yet, no rigorous justification can be found on the basis of the equations of motion of fluid flows.

Finding plausible closures, and possibly general ones covering a class of fluid flows as wide as possible, is a very active sector of fluid-dynamics research, commonly referred to as to **"turbulence modelling"** [9].

1.3 The role of geometry

Turbulent flows are **open** systems whose dynamic behaviour is highly dependent on the environment where fluid motion takes place. The environment influence is basically felt in terms of geometrical constraints to the fluid motion: most flows of practical relevance "spend their life" in bounded domains, such as ducts, pipes, combustion chambers, and other devices whose geometry is often quite complex.

As for any physical system described in terms of partial differential equations, the interaction between fluid flows and their environment is mathematically reflected by the type of boundary conditions to be imposed to the Navier-Stokes equations.

These boundary conditions pick-up specific solutions out of a whole class

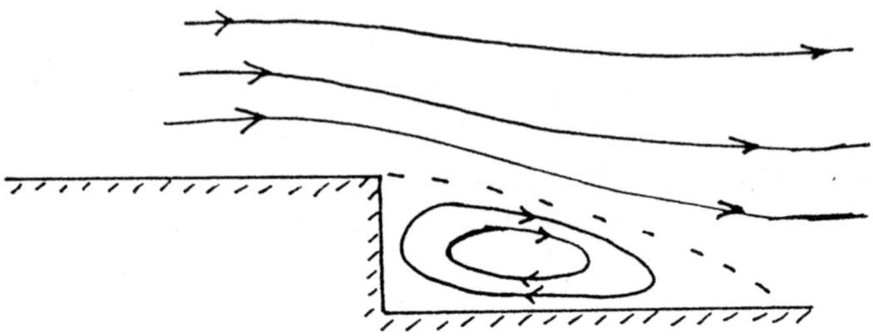

Figure 1.3: Backward facing step flow

of possible ones and ultimately dictate the large-scale feature of the fluid flow. This forms the rationale behind the effort of trying to classify the types of flow patterns which arise in connection with specific geometries, chosen in such a way as to obtain the best compromise between simplicity and fidelity [10].

Typical cases in point are channel flows, backward facing step flows, flow in a driven cavity (Fig. 1.3).

The hope is, of course, that the flow configuration generated by more complex real-life geometries, can be sensibly analyzed in terms of the "elementary" flow patterns associated with simpler but representative geometries [18].

A general remark is that the Navier-Stokes equations possess a number of discrete and continuous symmetries: spatial parity, Galilean invariance, rotational invariance, (isotropy) and, in the limit of vanishing viscosity, scaling invariance [11], i.e. invariance with respect to arbitrary deformations of the space and time scales.

These symmetries provide a powerful heuristics for the search of flow solutions, provided that the boundary conditions be also compatible with them.

This is precisely the point: realistic geometries are usually too complex in order for the aforementioned symmetries to be supported, with the result that flow solutions cannot be sought within pre-defined classes of solutions, but have to be found on a case-by-case basis.

So far, we've been presenting two major aspects which conspire to make fluid flows behaviour hard to predict: fluid turbulence and complex geometrical environment.

Fluid turbulence is responsible for generating a fully-coupled multiscale physics, while complex geometries are responsible for extensive symmetry breaking of the flow solutions.

There is a third aspect of fluid flow dynamics which sharply raises the complexity of this discipline: we refer to **reactive** flows, i.e. flows made up by several species which are allowed to transform one into another via chemical reactions.

1.4 Reactive Flows

Reactive Fluid Dynamics is a discipline of incredible richness and complexity, and forms the basis for a number of important applications, for instance combustion, which are gaining considerable attention, on account of increased awareness of environmental concerns [13] [14].

A comprehensive treatment of Reactive Fluid Dynamics goes obviously far beyond the scope of the present work. Here we shall content ourselves with a very few general considerations on the relationship between inert (non-reactive) and reactive flow dynamics, in order to gain a more complete view of the three-dimensional "TCG" (Turbulence-Chemistry-Geometry) space where flow complexity should be appreciated.

Flow reactivity can be summarized by two parameters: the amount of heat released per unit volume, ΔQ, and a typical time-scale over which this release occurs, τ_c. A reactive flow can be regarded to as a "minor perturbation" of an inert flow whenever the rate of temperature increase due to reactions $\dot{T}_{chem}$ is much smaller than the one produced by hydrodynamic effects (diffusion or convection) $\dot{T}_h$. Mathematically:

$$\frac{\dot{T}_{chem}}{\dot{T}_h} \simeq \frac{\Delta Q/\rho c_p \tau_c}{T/\tau_h} = \frac{\Delta T}{T}\frac{\tau_h}{\tau_c} \ll 1 \tag{1.9}$$

where ρ is the fluid density, c_p the specific heat, T the temperature and τ_h a typical hydrodynamic scale. The expression 1.9 is technically referred to as to Damkôhler number, Da, and can be taken as a qualitative measure of flow reactivity, pretty much the same way as the Reynolds number is a measure of flow turbulence.

From the equation 1.9 we see that significant departures from inert dynamics are to be expected for fast reactions ($\tau_c \ll \tau_h$) with non negligible heat release ($\Delta Q \simeq \rho c_p T$). Under these conditions, the flow temperature is no longer passively advected and diffused by the fluid flow, but it actively reacts back on it via the density changes imposed by the equation of state of the fluid.

Temperature activity, and its effect on the flow via density changes, is the hallmark of reacting versus inert flows.

Slow chemistry is seldom encountered in situations of practical interest: formation and destruction of pollutants in post-flame regions is one case. On the other hand, fast reacting flows occurr in the majority of practical applications, such as combustion in large furnaces and gas turbines, internal combustion engines, household burners, as well as reactive flows in the chemical industry.

1.5 The role of computers

The impressive advances of computer technology in the last two decades (see 1.4) have revolutioned the way how scientific research is conducted, both in industry and academia. Besides the two traditional avenues, theory and experimentation, computer advances have opened up a third dimension in the world of scientific investigation: numerical simulation. Numerical simulations are a pervasive presence across virtually all branches of scientific research, ranging from physics, chemistry, material science, to biology, economic sciences and many others. The idea is to gain new insights on unexplored frontieres of knowledge which are either unaccessible to direct experimentation and/or too complex to be investigated by means of analytical methods.

In the industrial context, computer simulation is playing a major role as complement/alternative to expensive laboratory experimentation to enhance the productivity by cutting the time-to-market and production costs.

Fluid-dynamics makes no exception to this rule: quite on the contrary, Computational Fluid Dynamics (CFD), the discipline spawned by the use of computers in Fluid Dynamics research, has gained a consolidated trailblazing reputation in the arena of computer simulation.

Proof is that most of the so-called "Grand-Challenges" in High Performance computing do involve fluid-dynamics issues, weather forecasting, aircraft design, combustion modeling being some examples in point. It is nowadays widely acknoweldged that in order to face the grand-challenges, greater computer resources will be required, both in memory and speed, with respect to those presently available. The targeted region of the computational spectrum ranges approximately from the Teraflop (10^{12} floating point operations per second) up to the Exaflop (10^{18} flop/s) regime [15].

Parallel computers are the most credited candidates to meet these needs. The key issue to be resolved with parallel computers is **scalability**, namely the ability to draw performance gains in linear proportion to the number of processors for a wide range of processors and applications. Achieving scalability is by no means a trivial task, and calls for significant advances and cooperation among at least three distinct scientific disciplines: computer science, numerical mathematics and physical modeling. The role of computer science is twofold: on the technological side, new parallel architectures geared to achieve optimal concurrency among the individual processors are continuously being explored.

As of today (mid 90's) several commercial machines in this breed have already come into existence [16].

Hardware is by no means the whole story: coordinating the concurrent operation of several processors (hundreds or thousands of) is a complex task which requires fairly sophisticated software tools, such as parallel compilers, debuggers, communication libraries [17]. These software tools must offer at a time high efficiency and-even more importantly- a high degree of conviviality, the latter being instrumental to the penetration of parallel computing into the industrial realm.

The role of numerical mathematics is quite obvious. The search of new algorithms and computational techniques characterized by an ever increasing exposure to parallel architectures lies at the very heart of Computational Fluid Dynamics.

Notwithstanding with their crucial role, computer science and numerical math alone will never be a full substitute for deeper physical insight into the physics of fluids. The modeling of fluid turbulence being perhaps the most striking example. This is why physical modeling will always represent a strategic component of CFD. New physical insights are primarily meant to produce better and more accurate models of fluid behaviour. Besides this direct output, an enlarged physical perspective of fluid dynamics may also lead to new mathematical formulations not necessarily based upon the Navier-Stokes equation of continuum mechanics.

This makes contact with the issue of numerical methods, which, in turn are intimately connected with computer science themes. As a result, we come somehow full circle back.

Covering the above matter within a single book is a daring task that goes way beyond the scope of this work. Here, we shall keep centered about basic methods of CFD and their amenability to present-day (distributed-memory) parallel computers.

The book is therefore primarily meant to serve as an introductory text for a non-specialist audience. For this very reason, only basic material will be presented, while more advanced and specialized topics such as adaptive and multigrid methods, have been deliberately omitted. It is hoped nonetheless that, despite its elementary nature, this work may provide a useful background also for more advanced readers.

The book is organized as follows.

In chapter I we have presented a brief account of the main themes behind CFD research. Chapter II deals with a synthetic reminder of the basic equations of fluid dynamics. In Chapter III we review the basics of numerical methods based upon grid discretization. In Chapter IV we introduce the fundamental ideas behind particle methods. Chapter V deals with a self-contained introduction to the guiding concepts of parallel computing. Chapter VI and VII deal with the parallelization of two important classes of numerical methods, explicit and implicit respectively.

Finally, in Chap VIII we conclude by presenting a selected sample of parallel CFD applications.

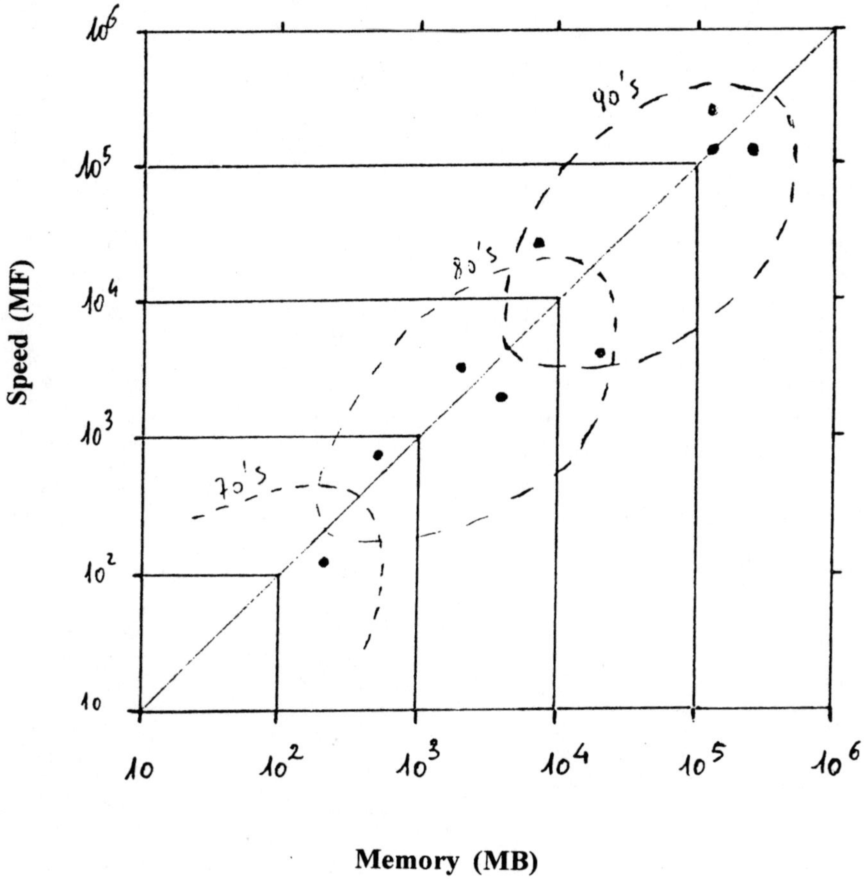

Figure 1.4: Advances in computer technology

Chapter 2

The basic equations of fluid flow

Taking the perspective of Classical Statistical Mechanics, fluids lie at the bottom level of a hierarchy of four, known as BBGKY hierarchy (Born, Bogoliubov, Green, Kirkwood, Yvon) [6] (Fig. 2.1).

At the top level, one finds the fully-atomistic description of a fluid as an ensemble of N ($N \simeq 10^{23}$) particles interacting according to the laws of Newtonian mechanics:

$$\begin{aligned} \dot{\vec{x}}_i &= \vec{v}_i \\ \dot{\vec{v}}_i &= \frac{\vec{F}_i}{m}(x_1 \dots x_N); \quad i = 1, N \end{aligned} \tag{2.1}$$

Obviously, due to the enormous number of degrees of freedom involved, there is no point of attempting a quantitative description of a macroscopic fluid flow by means of atomistic dynamics. Fortunately, besides being unpractical, such an atomic description is also needless because most of the details associated with equation 2.1 are completely irrelevant to the the purpose of a macroscopic description of fluid dynamics. This simple consideration suggests the resort to a statistical description of the fluid flow.

In three dimensions, the equation 2.1 corresponds to 6N degrees of freedom, which are most conveniently represented as a single point $P = (\vec{x_1}, \vec{x_2}, \dots \vec{x_N}, \vec{v_1} \dots \vec{v_N})$ moving about in a 6N-dimensional phase-space Γ.

Upon the assumption of ergodicity, i.e., any subregion $\Delta\Gamma$ of Γ having the same probability of being visited by the trajectory P(t), the system of

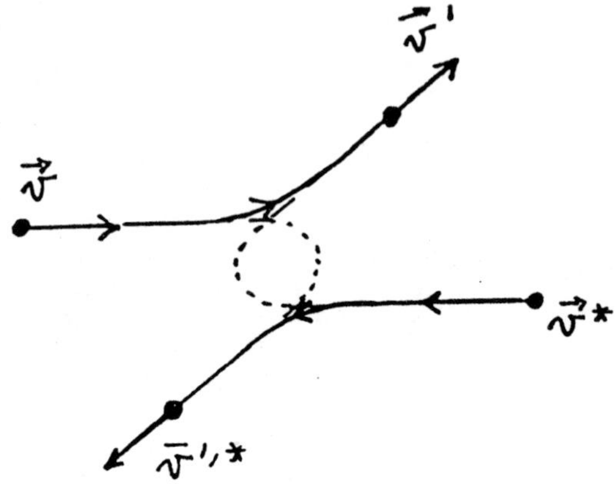

Figure 2.1: The four levels of the BBGKY hierarchy

Newtonian equations can be replaced by the Liouville evolution equation:

$$\frac{Df_N}{Dt} = \frac{\partial}{\partial t} f_N + \dot{\vec{x}}_i \cdot \frac{\partial}{\partial \vec{x}_i} f_N + \dot{\vec{v}}_i \cdot \frac{\partial}{\partial \vec{v}_i} f_N = 0 \tag{2.2}$$

where $f_N(\vec{x_1}, ...\vec{x_N}, \vec{v_1}...\vec{v_N}, t)\Delta\Gamma$ represents the probability of finding P(t) within $\Delta\Gamma$ at time t.

The Liouville equation represents the second level of the BBGKY hierarchy, and has the formal structure of an ideal fluid moving in the 6N dimensional space Γ.

Apparently, the Liouville equation doesn't represent any step forward with respect to the Newtonian formulation, since solving the equation 2.2 is just as difficult as solving equation 2.1.

The advantage, though, is apparent at a conceptual level, because the Liouville formulation opens the way to a systematic procedure of bottom-down "information removal" which ends up with the fluid equations of motion, thus bridging the gap between the atomistic and the macroscopic world.

The first step of this procedure is to recognize that, to the purpose of a macroscopic description, we don't need the full N-body distribution function f_N, but low-order ($N \simeq 1, 2$) functions are often perfectly adequate.

Redundant degrees of freedom can therefore be integrated out, down to the point where only f_1 and f_2 are involved. Under the assumption of

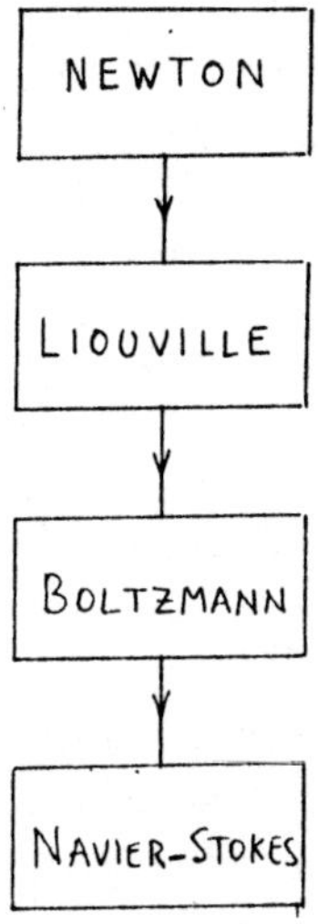

Figure 2.2: Sketch of a short-ranged binary collision among molecules

molecular chaos (no correlations between particles entering a collision), a self-contained equation for f_1 is obtained in the form (the suffix 1 as well as vector notation is omitted for brevity):

$$\frac{\partial f}{\partial t} + v\frac{\partial}{\partial x}f + \frac{F}{m}\frac{\partial}{\partial v}f = C(f,f) \tag{2.3}$$

where $C(f,f)$ represents short-range binary collisions.

Mathematically, the collision term involves integration over velocity space of quadratic non-linear terms [19], [20]:

$$C(f,f) = \int [f(v)f(v_*) - f(v')f(v'_*)]b\sigma(g)bdbdv_* \tag{2.4}$$

where unprimed (primed) speeds denote pre(post) collision state, g is the relative interparticle speed, σ the collision cross-section and b the impact parameter (Figure 2.2).

This is the celebrated Boltzmann equation, which lies at the heart of kinetic theory, i.e., the third level of the BBGKY hierarchy. The fluid level is ultimately attained by moving one step down upon integration over the velocity space, to derive dynamic equations for the fluid density,

$$\rho(x,t) = \int f dv$$

speed

$$u(x,t) = \int v f dv/\rho$$

and all the other quantities of macroscopic interest.

Integration upon $\vec{v}$ alone, is not enough because the number of unknows generated exceeds by one the number of available equations. This closure problem is usually circumvented by postulating small departures of $f(x,v,t)$ from local maxwellian equilibria parametrized by smoothly varying thermodynamics fields.

Here, "smallness" and "smoothness" are measured in terms of the Knudsen number, i.e. the ratio of particle mean free path, l_{mfp}, versus typical variation scale of macroscopic quantities, l_M: $Kn = l_{mfp}/l_M \ll 1$.

Although there's no question that fluids -as physical systems- definitely belong to the fourth level, from the methodological point of view, it may be often very convenient to describe them in terms of ensemble of "computational particles". Indeed, the mathematical techniques developed for the upper-lying BBGKY levels can prove quite useful and well suited to the modern computer architectures.

2.1 The Navier-Stokes equations

Having placed fluids in their proper perspective within the general framework of Statistical Mechanics, we now proceed to the presentation of the actual mathematical form of fluid equations [21].

To this purpose, the most intuitive and straightforward way is to formulate the principles of classical mechanics for the conservation of mass, momentum and energy, as applied to an infinitesimal [1] volume of the flow. In the absence of any external source, these read as follows:

From **mass conservation** , we obtain

$$\frac{\partial \rho}{\partial t} + \mathrm{div}(\rho \vec{u}) = 0 \tag{2.5}$$

where ρ is the fluid density and $\vec{u}$ the fluid speed.

[1] Infinitesimal means much smaller than any macroscopic volume of interest. This volume, though, must be significantly larger than the single-particle volume l^3_{mfp} in order to contain a sufficient number of equilibrated particles.

Momentum conservation yields

$$\frac{\partial}{\partial t}\rho\vec{u} + \mathrm{div}(\rho\vec{u}\vec{u}) = \mathrm{div}\ \overset{\Rightarrow}{S} - \vec{\nabla}p \tag{2.6}$$

where $\overset{\Rightarrow}{S}$ is the stress tensor.

Finally, **energy conservation** leads to:

$$\frac{\partial}{\partial t}\rho e + \mathrm{div}(\rho\vec{u}e) = \mathrm{div}(\overset{\Rightarrow}{S}: \vec{\nabla}\vec{u}) - \mathrm{div}(\vec{q}) \tag{2.7}$$

where e is the energy density (internal plus kinetic: $e = 1/2u^2 + W$) and $\vec{q}$ is the heat flux.

In order for equations 2.5-2.7 to form a closed set, the quantities p, $\overset{\Rightarrow}{S}, \vec{q}$ need to be expressed in terms of the fields characterizing the state of the flow, i.e. $\rho, \vec{u}, e$.

To this purpose, constitutive relations are introduced, which relate the shear stress tensor $\overset{\Rightarrow}{S}$ to the rate of strain tensor $\overset{\Rightarrow}{\gamma} = (\vec{\nabla}\vec{u} + (\vec{\nabla}\vec{u})^T)/2$ [2], and the heat flux $\vec{q}$ to the temperature gradient $\vec{\nabla}T$.

The constitutive relations take the following form:

$$\overset{\Rightarrow}{S} = \lambda\,(\mathrm{div}\,\vec{u})\ \overset{\Rightarrow}{I} + 2\mu\ \overset{\Rightarrow}{\gamma} - p\ \overset{\Rightarrow}{I} \tag{2.8}$$

$$\vec{q} = -K\vec{\nabla}T \tag{2.9}$$

where p is the fluid pressure, λ, μ are dynamic viscosity coefficients (known as Lamé coefficients), K is the thermal conductivity and $\overset{\Rightarrow}{I}$ is the unit tensor.

The equations 2.8 2.9 known also as "Navier-Stokes-Fourier" laws, rest upon a phenomenological assumption of weak departure from local thermodynamic equilibrium. Fortunately, this assumption seems pretty plausible for most fluids of practical relevance.

There are however important exceptions, a notable one being represented by "non-Newtonian" fluids, for which μ is a function of the stress tensor itself. A typical example of non-Newtonian fluid is ketch-up, oil and many polymer fluids for which μ also depends on the rate of change $d\overset{\Rightarrow}{S}/dt$, thus bringing memory effects into the game.

---

[2] The notation $(\vec{\nabla}\vec{u})^T$ means transpose. As a result $\overset{\Rightarrow}{\gamma}$ is equivalent to $\gamma_{ij} = 1/2(\partial_i u_j + \partial_j u_i)$, $i,j = 1,3$, $\partial_i \equiv \frac{\partial}{\partial x_i}$

The transport coefficients λ, μ, K generally depend on the local thermodynamic state (p, T), but in most conditions only the temperature dependence

$$\lambda = \lambda(T) \; ; \mu = \mu(T) \; ; K = K(T) \tag{2.10}$$

needs to be accounted for. In addition, due to the second law of thermodynamics, the coefficients λ, μ have to fulfill the following inequality:

$$3\lambda + 2\mu \geq 0 \; ; (\mu > 0) \tag{2.11}$$

In the absence of strong departures from local thermodynamical equilibrium, the Stokes relationship

$$3\lambda + 2\mu = 0 \tag{2.12}$$

is generally accepted as a valid approximation, so that any given fluid can be identified in terms of the two transport coefficients μ and K.

The last step to accomplish the closure of the equations 2.5- 2.7 consists of specifying an **equation of state** linking the fluid pressure to its density ρ and temperature T:

$$p = f(\rho, T) \tag{2.13}$$

as well as the functional dependence of the internal energy density W on the fluid temperature and pressure.

$$W = g(p, T) \tag{2.14}$$

The equations 2.13-2.14 take an especially simple form in the case of ideal gases (point-like particles with no potential energy) :

$$\frac{p}{\rho RT} = 1 \tag{2.15}$$

$$W = c_p T \tag{2.16}$$

where R is the universal gas constant (8.3143 J/mole K) and c_p is the specific heat of the gas at a constant pressure.

Deviations from ideality can be accounted for by replacing the right-hand side of 2.15 with the compressibility factor Z, which, accordingly to Van der Waals principle of corresponding states, is a universal function of the reduced temperature $T_r = T/T_c$ and pressure $p_r = p/p_c$, where T_c, p_c are the critical temperature and pressure of the given fluid (gas or liquid).

2.2 Various forms of the Navier-Stokes-Fourier equations

The Navier-Stokes equations can be written with respect to either a fixed reference frame (Eulerian form) or with respect to a frame moving with the local fluid speed (Lagrangian form) (see Figure 2.3) [22], [27].

2.2.1 Eulerian form

By using the relations 2.8-2.9, the conservation laws equations 2.5-2.7 take the following elegant and coincise conservative form for a set of 5 "generalized" densities ρ_α and currents $\vec{J_\alpha}$:

$$\partial_t \rho_\alpha + \text{div}(\vec{J}_\alpha^{(R)} + \vec{J}_\alpha^{(I)}) = 0 \; ; \alpha = 0, 4 \qquad (2.17)$$

with the following identifications:

$$\begin{array}{lll} \rho_0 = \rho; & \vec{J}_0^{(R)} = \rho\vec{u}; & \vec{J}_0^{(I)} = 0 \\ \rho_1 = \rho u_x; & \vec{J}_1^{(R)} = \rho\vec{u}u_x + p\hat{x}; & \vec{J}_1^{(I)} = \vec{\sigma}_x \\ \rho_2 = \rho u_y; & \vec{J}_2^{(R)} = \rho\vec{u}u_y + p\hat{y}; & \vec{J}_2^{(I)} = \vec{\sigma}_y \\ \rho_3 = \rho u_z; & \vec{J}_3^{(R)} = \rho\vec{u}u_z + p\hat{z}; & \vec{J}_3^{(I)} = \vec{\sigma}_z \\ \rho_4 = \rho e; & \vec{J}_4^{(R)} = \rho\vec{u}e + p\vec{u}; & \vec{J}_4^{(I)} = \stackrel{\Rightarrow}{\sigma} \cdot \vec{u} - \vec{q} \end{array} \qquad (2.18)$$

Here $\stackrel{\Rightarrow}{\sigma}$ is the viscous stress tensor, defined as:

$$\stackrel{\Rightarrow}{\sigma} = \stackrel{\Rightarrow}{S} + p \stackrel{\Rightarrow}{I} = \mu(\stackrel{\Rightarrow}{\gamma} - \frac{1}{3}\text{div}\,\vec{u} \stackrel{\Rightarrow}{I}) \qquad (2.19)$$

where use of the Stokes relation, equation 2.12 has been made. The notation $u_x, \vec{\sigma_x}$ stands for $\vec{u} \cdot \hat{x} \stackrel{\Rightarrow}{\sigma} \cdot \hat{x}$, respectively, $\hat{x}$ being the unit vector along the x-axis, and similarly for the y, z directions.

The equation 2.17 represent the Eulerian-conservative form of the NSF equations, the view of an observer sitting on a fixed volume of the fluid flow and detecting a rate of change in time of the physical quantity ρ_α matching exactly the flux of the current $\vec{J_\alpha}$ across the closed boundaries of the volume itself.

In the equation 2.17 we have deliberately split the current $\vec{J_\alpha}$ into two separate components $\vec{J}_\alpha^{(R)}$ and $\vec{J}_\alpha^{(I)}$ which stand for "reversible" and "irreversible" respectively.

The physical nature, and role on the flow dynamics, of the two components is profoundly distinct: reversible currents are energy conserving and therefore they are a faithful mirror of Newtonian mechanics as applied to a macroscopic volume.

Irreversible currents, on the contrary, **do** dissipate energy and momentum for, as a result of the quasi-local thermal equilibrium assumption, equations 2.8-2.9, they introduce thermodynamic forces which have no counterpart at the microscopic level.

Mathematically, this reflects in reversible currents involving only first-order space derivatives while irreversible currents require second order ones. First order derivatives imply free-streaming (hyperbolic) motion which is reversible because any initial condition can be recovered from an evolved one by merely inverting the sign of the speed. On the other hand, second order derivatives imply diffusive (parabolic) motion in which the initial state can only be regained from an evolved one by inverting the sign of the diffusion coefficient. Now, while negative speeds are perfectly acceptable in the physical world, negative diffusion is not, since it leads to gradient enhancement rather than smoothing as required by the second law of thermodynamics.

Notwithstanding with this general remark, it's however worth mentioning that negative "effective" transport coefficients can indeed show up during fluid motion, as a result of local instabilities leading to self-organized patterns. Cellular flames in turbulent combustion are a typical example of this kind of behaviour.

In a way, we may say that $\vec{J}_\alpha^{(R)}$ are in charge of letting fluid flows fulfill the first principle of thermodynamics, while $\vec{J}_\alpha^{(I)}$ account for the second one.

This leads to a straigthforward definition of a **perfect fluid**, as the one for which irreversible currents identically vanish:

$$\vec{J}_\alpha^{(I)} \to 0 \tag{2.20}$$

In view of the relations equations 2.8-2.9, the above condition is tantamount to erasing all transport phenomena by setting

$$\nu, K \to 0 \tag{2.21}$$

The limit of inviscid flow, equation 2.21, is very relevant and one that can be used in many practical applications, for instance, external flows around slanted bodies in aerodynamics.

However, it has to be handled with care because a quantitative assessment upon inviscidness cannot depend on the sheer value of the transport coefficients alone, equation 2.21, but has to take into account the relative weight of dissipative effects versus non-dissipative ones.

Hence, the correct criterion is rather

$$\frac{\vec{J}_{\alpha}^{(I)}}{\vec{J}_{\alpha}^{(R)}} \ll 1 \tag{2.22}$$

The subtlety is that the above ratio depends on the local flow configuration and is generally all but uniform across the flow.

Typically, dissipative effects tend to concentrate close to rigid walls where strong gradients, (boundary layers) develop. As a result, even if ν itself is very small, the term J^I can dominate the dynamics of the boundary layer.

By taking $J^{(I)} \simeq \nu \Delta u$ and $J^{(R)} \simeq (\vec{u} \cdot \vec{\nabla})\vec{u}$, equation 2.22 yields

$$Re = \frac{uL}{\nu} \gg 1 \tag{2.23}$$

where Re is the Reynolds number of the flow.

The limit of inviscid flow can therefore be characterized as the asympotic regime defined by $Re \to \infty$. Since $1/Re$ prefactors the higher order derivative in the Navier-Stokes equation, this limit is a singular one and has to be handled with great care.

2.2.2 The Lagrangian formulation

Using the identity

$$\mathrm{div}(\phi \vec{u}) = \phi \, \mathrm{div}\, \vec{u} + \vec{u} \cdot \vec{\nabla}\phi \tag{2.24}$$

where ϕ is any (smooth) scalar function, the system of equations 2.17, is readily turned into the following **non-conservative** form:

$$\begin{cases} \frac{D\rho}{Dt} = -\rho \, \mathrm{div}\, \vec{u} \\ \frac{D(\vec{u}\rho)}{Dt} = -\rho\, \vec{u} \, \mathrm{div}\, \vec{u} - \vec{\nabla}p + \mathrm{div}\, \vec{\vec{\sigma}} \\ \frac{D(e\rho)}{Dt} = -\rho\, e \, \mathrm{div}\vec{u} - \vec{u} \cdot \vec{\nabla}p + \mathrm{div}(\vec{\vec{\sigma}} \cdot \vec{u}) - \vec{\nabla}q \end{cases} \tag{2.25}$$

where $D/Dt = \partial/\partial t + \vec{u} \cdot \vec{\nabla}$ is the Lagrangian derivative taken along the fluid stream-lines ("material derivative").

These equations represent the rate of change of $\rho, \vec{u}, e$ as detected by an observer moving along with a fluid element at speed $\vec{u}$.

The most appealing feature of Lagrangian methods is a drastic simplification of the convective terms. In a local frame moving with the fluid the material derivative $D/Dt = \partial/\partial t + \vec{u} \cdot \vec{\partial}/\partial \vec{x}$ reduces to a simple one-dimensional derivative $\partial/\partial\tau$ with respect to the time τ parametrizing the characteristic of fluid motion (streamlines):

$$\frac{d\vec{x}(\tau)}{d\tau} = \vec{u}$$

The emerging picture is that of a fluid flow as a collection of particles sliding freely along the characteristics $\equiv$ streamlines, and interacting across them via pressure-waves and cross-diffusion.

The streamwise dependence reflects the hyperbolic component of fluid motion, while the cross-flow dependence reflects the elliptic one.

Lagrangian methods are well suited to achieve high accuracy for multi-dimensional super/hypersonic flows involving sharp shock-fronts. A similar consideration applies to high-Reynolds multiphase flows featuring moving interphase boundaries. In fact, since convective derivatives have been ruled out, positivity of mass and energy densities does not require artificial profile smoothing, i.e. numerical diffusion.

Lagrangian methods can profitably be combined with Eulerian methods to produce the so-called ALE (Arbitrary Lagrangian Eulerian) technique which is especially popular for combustion applications [25], [26].

Now, while it is true that mass, momentum and energy are conserved in the fluid element moving with the flow (making abstraction of the irreversible terms), the same is not true for the corresponding **densities** unless the volume itself is a conserved quantity.

To better appreciate this point, it's useful to remark that the equation 2.25 can also be rewritten as a mass conserving equation:

$$\frac{D\rho V}{Dt} = 0 \tag{2.26}$$

In fact, one has

$$\frac{D\rho V}{Dt} = \rho\frac{DV}{Dt} + V\frac{D\rho}{Dt} = 0 \tag{2.27}$$

whence

$$\frac{D\rho}{Dt} = -\rho\frac{1}{V}\frac{DV}{Dt} = -\rho \operatorname{div} \vec{u} \tag{2.28}$$

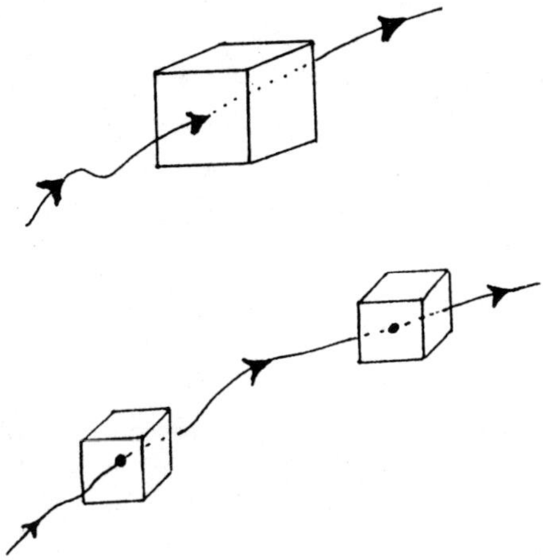

Figure 2.3: The Eulerian and Lagrangian formulations

where we remind that by definition div $\vec{u}$ is nothing but the relative rate of change in time of the moving volume [3]

$$\text{div}\vec{u} \equiv -\frac{1}{V}\frac{DV}{Dt} \tag{2.29}$$

From this discussion, it is clear that the non-conservative terms on the r.h.s. of equations 2.25, represent the effect of fluid **compressibility**. Whenever compressibility effects can be neglected, as for most liquids under ordinary conditions, the fluid density is trivialized to a constant value, and the equation 2.25 is actually replaced by the divergence-free condition

$$\text{div}\,\vec{u} = 0 \tag{2.30}$$

In practice, equation 2.30 corresponds to assuming $Ma \equiv u/c_s \to 0$, where Ma is the Mach number and c_s the sound speed.

This highlights the fact that the velocity field of an incompressible flow is divergence-free or, borrowing from electrodynamics, **solenoidal.**

The solenoidality condition allows significant simplifications in the fluid equations, which can be explicitly rewritten as

$$\frac{D\vec{u}}{Dt} = -\vec{\nabla}p + \nu\Delta\vec{u} \tag{2.31}$$

[3]whence the very definition of "divergence"

$$\frac{De}{Dt} = -\vec{u} \cdot \vec{\nabla} p + 2\nu\Delta(\frac{u^2}{2}) - \chi\Delta T \tag{2.32}$$

where $\nu = \mu/\rho$ and $\chi = K/\rho$ are the kinetic viscosity and conductivity coefficients. Under these conditions, the energy equation 2.32 is most conveniently recast in terms of an evolution equation for the flow temperature.

By dotting 2.31 with $\vec{u}$ and then subtracting from 2.32, we obtain ($c_p = 1$):

$$\frac{DT}{Dt} = \chi\Delta T \tag{2.33}$$

Hence, the eq. 2.6 describing the evolution of an incompressible flow can be summarized as follows:

$$\begin{aligned} &\text{div}\vec{u} = 0 \\ &\tfrac{D\vec{u}}{Dt} = -\vec{\nabla} p + \nu\Delta\vec{u} \\ &\tfrac{DT}{Dt} = \chi\Delta T \end{aligned} \tag{2.34}$$

The eq. 2.33 indicates that in an incompressible inert flow the temperature behaves like a passive scalar, i.e. is convected and diffused by the fluid flow, with no feedback reaction.

2.2.3 The vorticity-streamfunction formulation

Both Eulerian and Lagrangian formulations are based on so called "primitive variables", i.e. quantities such as density, momentum and energy, which bear an immediate physical meaning.

As is often the case in field theory, it may prove convenient to describe the field dynamics in terms of non-primitive, "derived", variables which are most suited to capture specific aspects and subtleties of the system dynamics.

In the case of fluid flows, an important derived variable is the fluid vorticity, defined as the degree of local rotation. Mathematically it is expressed by the curl of the velocity field

$$\vec{\omega} = \vec{\nabla} \times \vec{u} \tag{2.35}$$

The role of vorticity is particularly important for high Reynolds low-dissipation flows [29].

To better appreciate this point, let us remind that under the effect of stress field, the flow develops a deformational state which is represented by the deformation tensor $\vec{\nabla}\vec{u}$.

This tensor can be decomposed into a symmetric and antisymmetric components

$$\vec{\vec{\gamma}} = \frac{1}{2}(\vec{\nabla}\vec{u} + (\vec{\nabla}\vec{u})^T)$$

and

$$\vec{\vec{\Omega}} = \frac{1}{2}(\vec{\nabla}\vec{u} - (\vec{\nabla}\vec{u})^T),$$

which reflect fluid deformation and rigid body rotation respectively.

The former is responsible for dissipation, while the latter expresses the ability of fluid flow to develop rotational motions virtually free of dissipation (after all, that's just the principle of the wheel!).

By self-contracting the deformation tensor, we obtain:

$$\vec{\nabla}\vec{u} : \vec{\nabla}\vec{u} = ||\gamma||^2 + ||\Omega||^2 \tag{2.36}$$

which shows that part of the work done on the flow goes into dissipation (irreversible component) and part gets stored in the form of rotational energy which can possibly be regained (reversible component).

The link with the vorticity is now immediate:

$$||\Omega||^2 = \omega^2 \tag{2.37}$$

as it results from the fact that Ω_{ij}, being antisymmetric, is characterized by only three independent components to be identified with the three cartesian components of vorticity vector:

$$\Omega_{xy} = \omega_z \; ; \Omega_{xz} = \omega_y \; ; \Omega_{yz} = \omega_x \tag{2.38}$$

The observation that vorticity seems to be the "relevant" field for low-dissipative flows, has led to the notion of a fluid flows as a collections of elementary "quanta" of vorticity named "vortons".

Besides its theoretical appeal, the idea of vortons has also spawned an original computational technique which will be discussed in the sequel of this work.

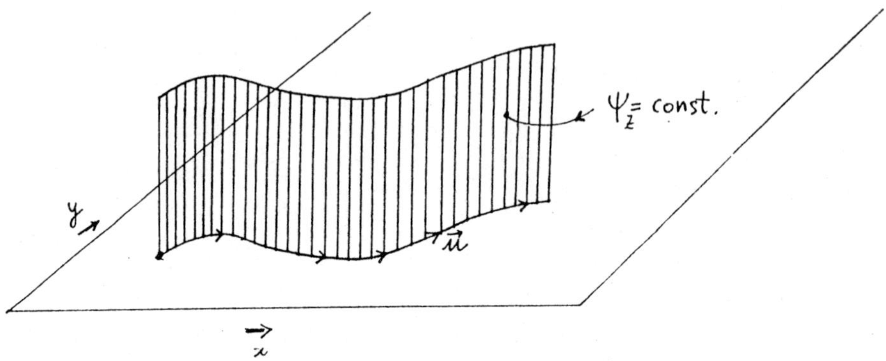

Figure 2.4: Streamlines of a two-dimensional flow, are defined as intersection of a $\psi_z = const.$ sheet with the plane where the flow takes place.

The notion of vorticity leads to especially elegant formulations for the case of incompressible flows. In fact, as is known from electrodynamics, a divergence-free field implies the existence of a vector potential $\vec{\psi}$ such that

$$\vec{u} = \vec{\nabla} \times \vec{\psi} \tag{2.39}$$

By combining eq. 2.39 and 2.35, we obtain

$$\vec{\omega} = \Delta \vec{\psi} \tag{2.40}$$

which shows that the vorticity plays the role of a vector "charge" for the potential $\vec{\psi}$: the elliptic nature of this equation is the result of the incompressibility assumption which implies infinite speed of propagation of density waves.

In the fluid-dynamic literature the vector potential $\vec{\psi}$ is known as "streamfunction" because (in two dimensions) it is easily proved that the flow streams along surfaces of constant $\psi \equiv \psi_z$. (Fig. 2.4)

The vorticity-streamfunction formulation of the NS equations is obtained by taking the curl of the second eq. 2.34. This yields:

$$\frac{D\vec{\omega}}{Dt} = -(\vec{\omega} \cdot \vec{\nabla})\vec{u} + \nu \Delta \vec{\omega} \tag{2.41}$$

to be complemented by eq. 2.39 and 2.40.

The eq. 2.41 shows that vorticity is not just simply advected and diffused like a passive scalar, but it is generally also produced/annihilated via the stretching term $(\vec{\omega} \cdot \vec{\nabla})\vec{u}$.

This term identically vanishes in two-dimensions; this is a source of major qualitative differences between the behaviour of two and three dimensional turbulent flows.

The $\omega - \psi$ formulation is particularly suited to field-theoretical treatments as well as to computational studies aimed at the understanding of fully developed homogeneous turbulence.

For more involved geometries, involving solid obstacles, a problem arises with the correct treatment of boundary conditions which need to be expressed in terms of the primitive variables $p, \vec{u}$ rather than $\vec{\omega}, \vec{\psi}$ [30].

2.2.4 Potential flows

In many flows of practical interest the vorticity is concentrated in well defined regions, while outside these regions the flow can plausibly be regarded as completely **irrotational**, i.e. $\vec{\omega} = 0$.

A typical case is the flow past a solid body, where the vorticity is confined within the recirculation region of the wake behind the body.

In view of eq. 2.35, an irrotational flow necessarily derives from a potential ϕ such that

$$\vec{u} = -\vec{\nabla}\phi \qquad (2.42)$$

If, in addition, the flow is also incompressible, combination of eq. 2.42 and 2.30 yields

$$\Delta\phi = 0 \qquad (2.43)$$

which shows that the potential is harmonic.

Potential flow solutions are of special value for external aerodynamics calculations in which the irrotational flow in the exterior region is coupled to interior boundary layers by means of matching techniques.

Chapter 3

Numerical solution of fluid flow equations: Grid Methods

The numerical techniques for the solution of fluid flow equations split into two broad categories:

- Grid methods (GM)
- Particle methods (PM)

Grid methods will be discussed according to the following classification

- Local methods (Finite Differences, Volumes, Elements)
- Global methods (Spectral method)

The particle methods considered here are the following

- Molecular Dynamics
- Direct Simulation Montecarlo
- Lattice Gas and Lattice Boltzmann
- Vortex method

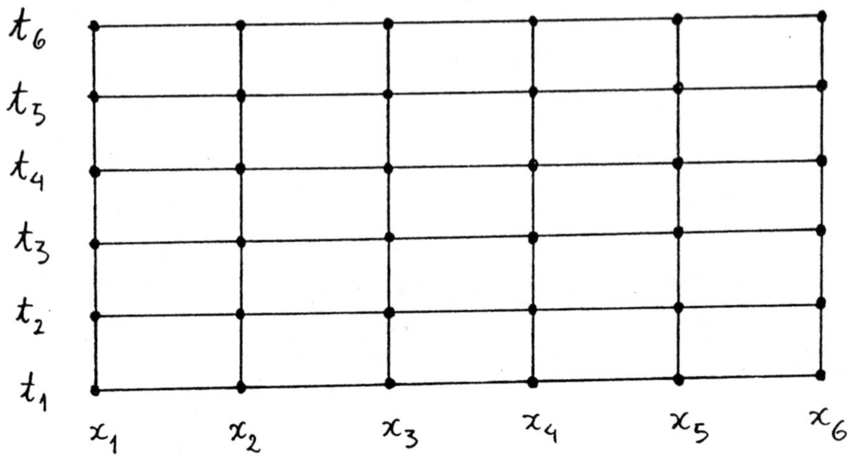

Figure 3.1: A discrete space-time grid

This choice is by no means exhaustive of the impressive spectrum of modern CFD methodolgies, nor does it reflect any author's assumption on their relative importance. It simply responds to the need of illustrating a number of basic concepts with a minimal - and yet representative - amount of information.

3.1 Grid methods vs. Particle methods

Within GM, the space-time continuum $C = \{\vec{x}, t\}$ is replaced by a **discrete** grid (Fig. 3.1)

$$G_{LN} = \{(\vec{x}_l, t_n)\,; l = 1, L, n = 1, N\} \tag{3.1}$$

and the continuous functions characterizing the fluid state, $\{\rho(\vec{x}, t), \vec{u}(\vec{x}, t), ...\}$, are represented by a corresponding set of discrete sequences $\{\rho_{ln}, \vec{u}_{ln}, ...\}$ which may, but need **not**, represent the value of the fields in the discrete point $(\vec{x}_l, t_n)$, i.e. $\rho_{ln} = \rho(\vec{x}_l, t_n)$.

This discretization procedure, once applied to the flow equations, generates a set of algebraic equations for the unknows $\rho_{ln}, \vec{u}_{ln}, ...$, that are usually solved by linear algebra procedures.

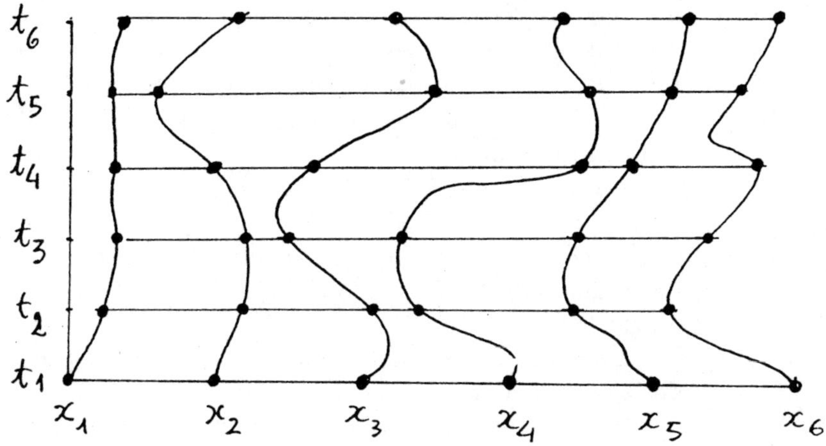

Figure 3.2: A discretized space-time bundle of trajectories

Within PM, the space-time continuum is not discretized by a grid, but rather with the aid of a discrete bundle of particle trajectories:

$$B_{N_P,N} = \{\vec{x}_p(t_n;\vec{x}_{0p})\ ; p = 1, N_P\ , n = 1, N\} \tag{3.2}$$

where $\vec{x}_p(t;\vec{x}_{0p})$ represents the trajectory of the p-th particle starting at $\vec{x}(t=0) = \vec{x}_{0p}$.

Once the particle trajectories are obtained by integrating in time the equations of motion, the fluid field can be derived by summing over the whole set of N_P particles and averaging over an ensemble of realizations obtained by changing the initial conditions:

$$\begin{aligned} \rho(\vec{x}_l, t_n) &=< \sum_{p=1}^{N_P} \phi(\vec{x}_l - \vec{x}_p(t_n)) > \\ u(\vec{x}_l, t_n) &=< \sum_{p=1}^{N_P} \dot{\vec{x}}_l \phi(\vec{x}_l - \vec{x}_p(t_n)) > \end{aligned} \tag{3.3}$$

where $\phi(\xi)$ is an appropriate shape-function peaked around $\xi = 0$, whose width measures the spatial size of a single particle. Brackets denote ensemble averaging.

In terms of the BBGKY hierarchy described in the previous chapter, it is clear that grid methods belong to the fourth-level, the fluid one, whereas particle methods belong to the first level, i.e. the Newtonian one. In this respect, going back to the atomistic level to describe macroscopic phenomena would seem, at a first glance, a bit of a nonsense. This is not so, as it becomes apparent once we realize that the particles involved by this procedure are just conceptual and computational entities which should by no means be identified with the real particles (atoms, molecules) constituting a real fluid. Otherwise, an astronomical number N_P would be required to describe any realistic macroscopic situation ($N_P \simeq 10^{23}$!).

Quite on the contrary, one of the basic lessons of modern Computational Statistical Mechanics, is that a limited set of "pseudo-particles", say $10^3 - 10^6$, is able to reproduce non-trivial fluid behaviour, such as the formation of convective cells in a fluid heated from below (Benard convection) or vortex shedding past a cylinder.

The ultimate reason for this fortunate instance is the fact that many details of the microscopic dynamics have no impact on the actual macroscopic dynamics because their cumulative effects cancel out along the chain which takes the first level of the BBGKY hierarchy down to the fourth one.

Loosely speaking, Grid Methods reflect the Eulerian point of view, while Particle Methods are a natural expression of the Lagrangian point of view. As a result, PM feature a sort of built-in form of concurrency (particle trajectories evolving simultaneously) which makes them increasingly competitive as the focus of high-performance computing is shifted towards parallel computers.

Whether they will ever come to the point where they can outperform the most consolidated GM is still an open question [31].

3.2 Grid methods

Denoting by $f(\vec{x}, t)$ any unknown function describing the fluid flow at point $\vec{x}$ and time t, the goal of grid-methods is to generate a consistent set of algebraic equations for the unknows f_{ln} , the discrete representation of the function $f(\vec{x}, t)$.

Grid methods split into two main categories, **local** and **global**, depending on whether the coefficients f_{ln} describe a property which is local to the grid point $\vec{x}_l$ at time t_n or rather a property which is global, i.e. distributed all over the spatial grid $X_L = \{\vec{x}_l \ , l = 1, L\}$ [32], [33].

Among the local methods, the most popular ones are:

- Finite Differences (FD)
- Finite Volumes (FV)

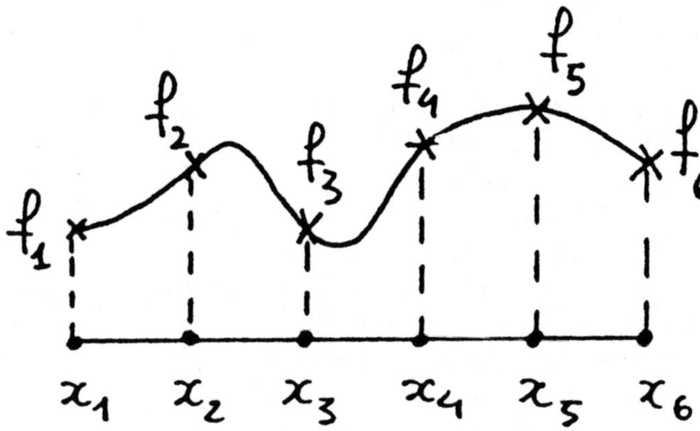

Figure 3.3: A finite-difference representation of the function $f(x)$.

- Finite Elements (FE)

while perhaps the most popular technique among the global ones is the Spectral Method (SM) [34].

A detailed description of any of these techniques would require a whole monograph in itself, and goes therefore far beyond the scope of this book. Here, we shall content ourselves with a cursory view of these methods aimed at highlighting their common features mostly in connection with their amenability to parallel computing.

3.2.1 Finite difference methods

The essence of the finite-difference method (FDM) consists of representing any unknown function $f(x)$ (for simplicity, we drop the time dependence as well as multidimensionality of the configuration space) by a set of discrete values $\{f_l;\ l = 1, L\}$, with the implicit identification [35]

$$f_l = f(x = x_l) \tag{3.4}$$

The eq. 3.4 doesn't have any bearing on the behaviour of $f(x)$ in between the discrete sequence x_l, and in particular doesn't impose any specific form for the discrete representation of first and second order spatial derivatives which appear in the Navier-Stokes equations.

The decision on how to turn these derivatives into a discrete form is therefore a matter of choice related to the desired accuracy, stability and other basic requirements of the numerical scheme. Perhaps, the simplest choice is the Euler-forward representation:

$$\left.\frac{df}{dx}\right|_{x=x_l} \rightarrow D^+ f|_{x=x_l} \equiv \frac{f_{l+1} - f_l}{\Delta x} \tag{3.5}$$

where $\Delta x = x_{l+1} - x_l$. This discrete differentation is simple but only **first order accurate**, in that the replacement eq. 3.5 is exact only if the function f varies linearly within the interval $[x_l \, , x_{l+1}]$.

The discretization 3.5 is first order accurate and generally introduces a **discretization error** of the order of Δx.

Symbolically:

$$\frac{d}{dx} = D^+ + O(\Delta x) \tag{3.6}$$

within the class of locally polynomial functions.

Accuracy can be improved but only at the expense of **locality**, i.e. by widening the neighborhood of the discrete differentiation.

For example, a popular choice ensuring 2nd order accuracy is provided by the central differences scheme (a uniform mesh is assumed here):

$$\left.\frac{df}{dx}\right|_{x=x_l} \rightarrow D_C f|_{x=x_l} = \left(\frac{D^+ + D^-}{2}\right) \equiv \frac{f_{l+1} - f_{l-1}}{2\Delta x} \tag{3.7}$$

The operator D_C differentiates exactly locally quadratic functions, but only at the price of a doubly reduced locality (2 neighbors of x_l instead of one only).

This tradeoff between locality and accuracy can be brought to the extreme, by introducing the following "exact" differentation [36]

$$\left.\frac{df}{dx}\right|_{x=x_l} \rightarrow D_E f = \sum_{m=1}^{L/2} W(m) \frac{f_{l+m} - f_{l-m}}{2\Delta x} \tag{3.8}$$

with

$$W(m) = (-1)^m \cdot 2 \cdot \frac{m\pi/L}{tg(m\pi/L)} \tag{3.9}$$

It is readily seen that D_E is nothing but a weighted average, with weight $W(m)$, of all possible centered schemes which can be defined in the lattice. Thus, D_E achieves exponential accuracy, but only at the price of a complete loss of locality.

The tradeoff is clear: higher order schemes offer appealing savings in terms of number of grid points required to achieve a given accuracy. On the other hand, this advantage is partially offset by the fact that each grid point demands more storage and CPU (central process unit) time to be updated.

Locality is highly prized on any computer architecture because it minimizes the data traffic among the different functional units of a given computer or, even more importantly, among different processors of a multicomputer parallel architecture.

For many applications, second order accuracy is often regarded as an acceptable compromise between accuracy and efficiency.

To sum up, the FD procedure leads to a set of ordinary differential equations of the following form:

$$\frac{df_l}{dt} = \sum_{m=1}^{L} A_{lm} f_m \tag{3.10}$$

where A_{lm} are the coupling coefficients stemming from the discretization of the differential operators.

Typically, the matrix A is **sparse**, i.e. only a few entries take non-vanishing values as the indices l, m lie sufficiently far apart. This reflects the local nature of the underlying differential operators.

$$A_{lm} \neq 0 \quad if |l - m| > B_W$$

Here B_W denotes the matrix **bandwidth**, a parameter that plays a crucial role on the efficiency of the solution to the eq. 3.10.

Finite Differences are quite straightforward and intuitive as long as the mesh geometry is simple enough, like in the case of cartesian geometries with boundaries lying on a mesh coordinate line.

Non-cartesian geometries can be handled by means of conformal transformations, mapping the physical domain onto a cartesian computational one. This technique is however constrained to the case where boundaries can be naturally fitted to a mesh coordinate and doesn't extend easily to more complex geometries for which no single **global** transformation with such a property exists.

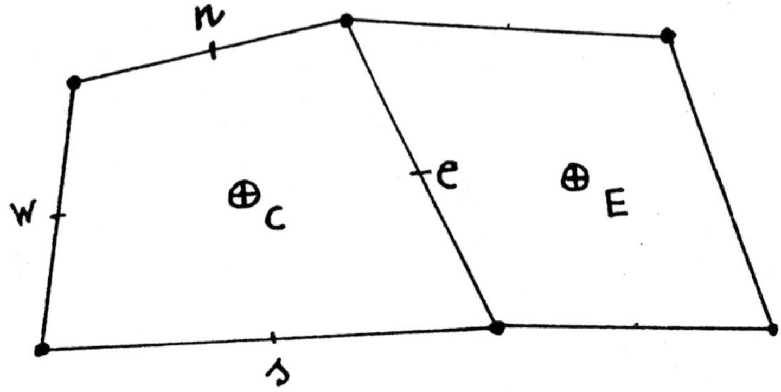

Figure 3.4: A typical quadrilater cell of the finite-volume mesh.

This problem is solved by moving from **coordinate-based** methods, like FD, to **element** or **cell**-based methods like Finite Volumes or Finite Elements.

3.2.2 The Finite Volume method

Much of the geometrical flexibility missing in the FD formulation can be gained by decomposing the geometrical domain into a series of non-overlapping elementary volumes (cells), typically exahedra in three dimensions and quadrilateral in two.

Each of these volumes is topologically equivalent to a cube, and so is the union of all the volumes. This means that there still exists a transformation taking the geometrical domain into a cubic *structured mesh* but this transformation is only piece-wise smooth, in the sense that the Jacobian of the transformation may undergo a discontinuity when passing from one element to another.

This loss of smoothness is the price to pay to achieve geometrical flexibility. Trading smoothness for flexibility is generally a good bargain, in that smoothness can be recovered by resorting to an **integral** formulation of the fluid equations.

To this purpose, let us consider the conservative form eq. 2.17 and apply

Gauss theorem to each individual volume Ω_C. This yields:

$$\frac{dF_C}{dt} + \phi_C = 0, \qquad C = 1, NC \tag{3.11}$$

where

$$F_C = \frac{1}{V}\int_{\Omega_C} \rho dV \tag{3.12}$$

is the total "mass in the volume Ω_C and

$$\phi_C = \frac{1}{V}\int_{\partial\Omega_C} \vec{J}\cdot d\vec{S} \tag{3.13}$$

is the flux across the surface of Ω_c.

The eq. 3.11 represents a set of NC ordinary differential equations for the NC unknowns F_C, which become consistent once a prescription is made on how to compute the surface integrals eq. 3.13 in terms of the cell unknowns F_C.

This calls for the choice of an interpolation scheme, i.e. a finite-difference procedure. This choice is by no means unique depending on the nature (convective, diffusive) of the flux term, the desired degree of accuracy, stability and other properties of the numerical scheme.

An important point, possibly the most basic asset of the FV method, is that no matter which interpolation scheme is chosen, the method remains **by construction** conservative because the flux leaving a cell across a given surface must necessarily match the flux entering the neighboring cell sharing that surface as a common boundary.

For instance, by labelling the cells as "e, n, w, s" for "east north west south" respectively, it is clear that $J_C^{east} = -J_E^{west}$ and similarly for all the remaining neighbors (see Fig. 3.4). As a result, by summing up all over the cells and assuming no-flux boundary conditions, we obtain

$$\frac{d}{dt}\sum_C F_C = -\sum_C \phi_C = 0 \tag{3.14}$$

This means that the total "mass" $F \equiv \sum_C F_C$ is **exactly** conserved (up to machine round-off) during the evolution.

A possible choice for first-oder space derivatives is the centered difference scheme, which reads as:

$$F_e = \alpha F_C + \beta F_E \tag{3.15}$$

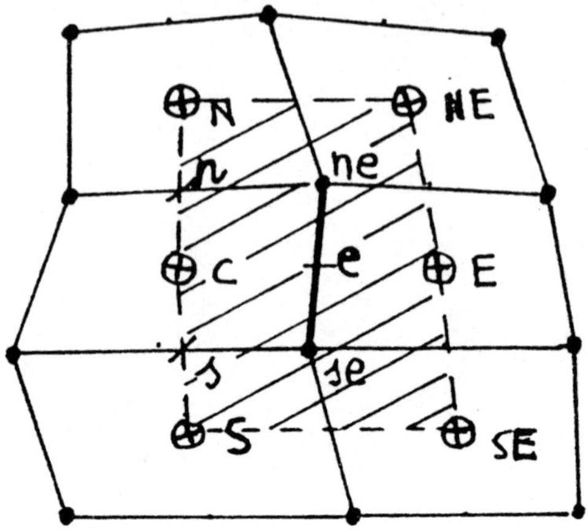

Figure 3.5: The dual cell Ω_e^* (dashed region) needed to evaluate "east" fluxes with respect to the nodal unknown sitting in C

with

$$\alpha = |Ce|/|CE|; \qquad \beta = |eE|/|CE| \tag{3.16}$$

and similar expressions for the other surfaces.
This leads to dynamic equations of the general form

$$\frac{dF_C}{dt} = \sum_{C'=1}^{NC} A_{CC'} F_{C'} \tag{3.17}$$

where $A_{CC'}$ is the (sparse) matrix representing the coupling of the actual unknown F_C to its neighbors $\{C' = E, N, W, S\}$ and itself ($C' = C$).

This system is the FV counterpart of the FD evolution equations 3.10.

It is important to remark that viscous flows imply the discretization of 2nd order derivatives, so that the interpolation of first order derivatives on the mesh edges is required. This adds to the complexity of the scheme. As an example, let us consider the FV discretization of the term $\vec{\nabla} u_i$, ($i = 1, 2, 3$) on the "east" surface contouring the cell centered in C (Fig. 3-5).

One possibility is to take a weighted average across the dual cell Ω_e^*, centered in e (see dashed region in Fig. 3.5)

$$\frac{1}{V_e^*}\left(\oint_{\Omega_e^*}\frac{\partial u_i}{\partial x}\vec{i}dxdy+\oint_{\Omega_e^*}\frac{\partial u_i}{\partial y}\vec{j}dxdy\right)=$$

$$\frac{1}{V_e^*}\left(\vec{i}\oint_{\partial\Omega_e^*}u_i\,dy+\vec{j}\oint_{\partial\Omega_e^*}u_i\,dx\right) \tag{3.18}$$

where V_e^* is the volume of the dual cell Ω_e^*.
A possible discretization of the line-integrals is:

$$\oint_{\partial\Omega_e^*}u_idy=u_{i,E}(y_{NE}-y_{SE})+u_{i,NE}(y_{NE}-y_N)+$$
$$u_{i,C}(y_N-y_S)+u_{i,SE}(y_{SE}-y_S) \tag{3.19}$$

and a similar expression for $\oint_{\partial\Omega_e^*}u_idx$.

By interpolating the edge values y_{NE}, y_N, y_S, y_{SE} and the nodal values y_{NE}, y_S, the eq. 3.19 can be finally expressed as a linear combination of the cell variables $\{u_C, u_S, u_{SE}, u_E, u_{NE}, u_N\}$.

The same procedure, repeated over the four surfaces "east", "north", "west", "south", generates an expression for the diffusion terms which involves the "next-to-nearest-neighbors" of cell C, i.e., a 9-point stencil, or "computational molecule": $nnn \equiv \{C\{E, NE, N, NW, W, SW, S, SE\}\}$.

The FV method affords a great deal of geometrical flexibility due to the capability of local mesh deformations which allow the numerical grid to adapt itself to complex and irregular geometries. Yet, there is a limit to this flexibility, i.e. to the degree of distorsion that can be tolerated without corrupting the numerical solution. Typical important indicators of mesh distorsion are:

- Degree of non-orthogonality $\delta = \alpha - \frac{\pi}{2}$
- Degree of cell-stretching $\epsilon = \frac{b}{a}$
- Local mesh curvature $\gamma = \beta - \pi$
- Cell face curvature (warp angle) w

whose geometrical meaning is clearly illustrated by Fig. 3-6

The construction of an optimal grid for a given geometry is a delicate and very important task, often constituting the most time-consuming stage

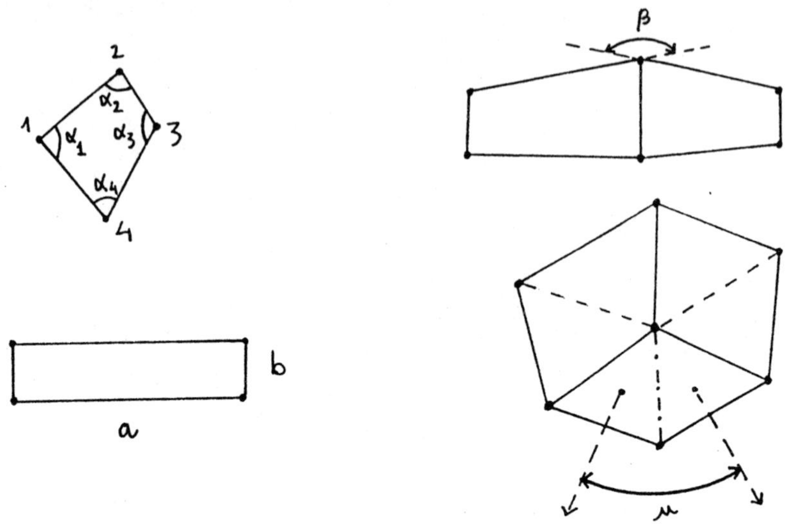

Figure 3.6: The main geometric indicators of mesh distortion in a finite volume grid

of a fluid-flow analysis. This is the reason why the issue of **automatic** grid-generation by means of specialized computer codes aimed at relieving the user from painful and lenghty manual intervention, has become a very active and strategic branch of modern CFD research [37].

The surplus of geometrical power of FV with respect to FD translates into an obvious additional workload from the computational point of view. This is the work associated with the **assembly** stage, i.e. the stage in which the coupling matrix $A_{CC'}$ is constructed starting from the geometry of the grid. Typically, this requires storing the following information

- Mesh-node coordinates $\vec{x}_\nu$, $\nu = 1, NN$
- Cell-face normal surface vectors $\vec{S}_\sigma$, $\sigma = 1, 3NC$
- Mesh-cell volumes V_c, $c = 1, NC$

where NN, NC is the number of nodes (grid points) and cells respectively. Straightforward vector-tensor algebra is then applied to obtain the coupling coefficients $A_{CC'}$.

The assembly-stage can take a substantial fraction of the CPU cost for those problems in which the matrix A needs to be reassembled many times within iteration loops dealing with the non-linearity of the convective terms of the Navier-Stokes equations.

In the last decade, the FV methods have gained an increasing popularity within the CFD community. So much so, that they have made their way into several commercial CFD packages.

3.2.3 The Finite Element method (FEM)

As we have seen in the previous section, the FV method achieves a great deal of geometrical flexibility over FD by relaxing the constraint of a single, global, geometrical transformation, in favor of a series of local coordinate transformations taking arbitrary hexaedra into logical cubes. The resulting discontinuities of the first-order derivatives are then smeared-out by moving to an integral formulation via Gauss theorem applied on a cell-by-cell basis.

This permits to solve the Navier-Stokes equations on fairly distorted grids, all of which retain, however, an elementary structure from the logical point of view, in the sense that each node of the mesh is uniquely identified by three indices, say, (i, j, k). In other words, the mesh is **structured**, i.e. topologically equivalent to a cube.

An important and practical property of structured meshes is that the number of neighbors (the "coordinations number" of the lattice), is a constant across the grid, so that the neighbors of any given node (i, j, k) are automatically identified as $(i \pm \delta_i, j \pm \delta_j, k \pm \delta_k)$ where $\delta_i, \delta_j, \delta_k$ -the displacements- are integers in the sequence $\{0, 1, 2, ..., NH\}$ where NH is the number of neighbors depending on the order of the interpolation method.

It is apparent that further geometrical flexibility can be gained by relaxing the constraint that

i) each point should be uniquely identified by a logical triplet (i, j, k).

ii) The number of neighbors NH be the same for each node of the grid

This points to "amorphous" or in CFD terminology **unstructured** grids.

Typical unstructured elements are triangles in 2-D and tethraedra in 3-D (Fig. 3.7):

From the practical point of view, the advantage of unstructured meshes is the possibility of **local mesh refinements**, i.e. clustering the grid points there where they are needed and rarefying them in regions where the flow-activity is weak.

The fundamental geometrical entities in an unstructured grid are **elements**, defined as a collection of **nodes** linked together by **links** defining the grid connectivity.

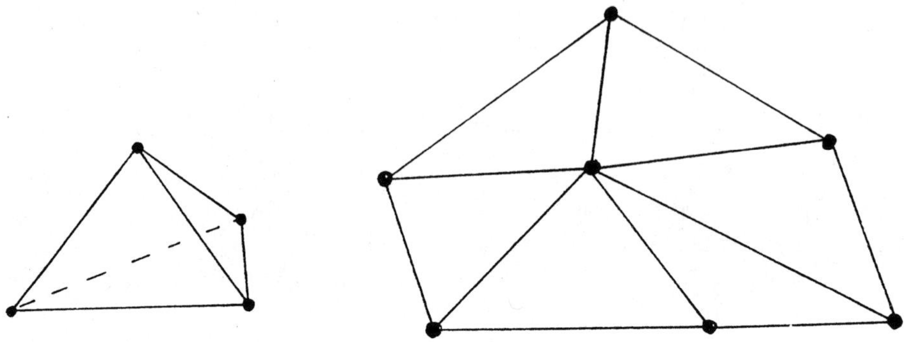

Figure 3.7: Triangles and tetrahedra as typical two and three-dimensional finite elements

In principle, one could still apply a FV procedure to each element; the problem is, however, that the interpolation procedure required to handle the surface integrals in terms of cell-centered quantities, would be somehow obscured by the lack of structure of the grid.

Still, efficient FV methods on unstructured grids have been developed in the recent years [38].

It is, however, much more convenient to focus one's attention on the nodes, and express the unknown functions in terms of their **nodal** values $f_\nu(t), \nu = 1, NN$.

Thus, a generic function f of space and time is represented in the following form [39], [40], [41]:

$$f(\vec{x}, t) = \sum_{\nu=1}^{NN} f_\nu(t)\psi_\nu(\vec{x}) \tag{3.20}$$

where $f_\nu(t)$ are the nodal unknowns, and $\psi_\nu(\vec{x})$ are appropriate shape-functions (**finite elements**) localized around the node $\vec{x}_\nu$.

The shape functions obey the "tent" property

$$\psi_\nu(\vec{x}_\mu) = \delta_{\nu\mu} \qquad \nu, \mu = 1, NN \tag{3.21}$$

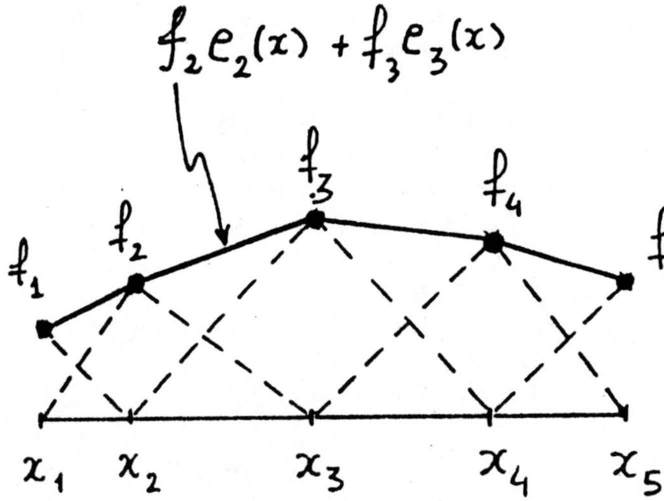

Figure 3.8: The finite-element representation of the function $f(x)$ with linear basis functions.

and take non-zero values only within the elements sharing the given node as a common vertex.

We use the expression "tent" property to convey the intuitive picture of the global function $f(\vec{x},t)$ as an envelope of localized support functions $\psi_\nu(x)$ which take the value 1 at $x = x_\nu$ and 0 at any other node.

As a result, the restriction to Ω_e of the function $f(\vec{x},t)$ can be expressed as the sum of all nodal contributions of the nodes belonging to Ω_e

$$f_e(\vec{x},t) = \sum_{\nu\in\Omega_e} f_\nu(t)\psi_\nu(\vec{x}) \tag{3.22}$$

For example, with reference to the figure 3.8, we have

$$\begin{aligned}\Omega_1 &= [x_0, x_1],\ f_1(x) = f_0\psi_0(x) + f_1\psi_1(x)\\ \Omega_2 &= [x_2, x_3],\ f_2(x) = f_2\psi_2(x) + f_3\psi_3(x)\end{aligned} \tag{3.23}$$

and the global solution results from the summation of the contributions over the entire element set:

$$f(\vec{x},t) = \sum_{e=1}^{NE} f_e(\vec{x},t) = \sum_{e=1}^{NE}\sum_{\nu\in\Omega_e} f_\nu\psi_\nu(\vec{x}) \tag{3.24}$$

In conclusion, the idea of the FE is to represent the unknown functions in terms of a set of interpolating piece-wise (local support) polynomial basis functions $\{\psi_\nu(x)\}$. Manifestly, by raising the order of the interpolating polynomial, increasing smoothness can be obtained.

For instance, by using 1st order "hat-functions" such as those represented in fig. 3.8, only inter-element continuity of the function is guaranteed, while its derivatives are generally discontinuous. Continuity of the function itself plus its first and second order derivatives is obtained by using suitably chosen third-order polynomials $S_3(x)$ (cubic-splines).

The equations of motion for the nodal amplitudes $f_\nu(t)$ are obtained by means of the Ritz-Galerkin weak-formulation.

For a generic evolution equation of the form

$$\partial_t f = \mathcal{L} f \tag{3.25}$$

the weak-formulation consists of projecting the eq. 3.25 onto an appropriate test function space $\mathcal{F}_T$, according to

$$(\phi, (\partial_t - \mathcal{L}) f) = 0 \tag{3.26}$$

where $(\cdot, \cdot)$ denotes scalar product in $\mathcal{F}_T$ and ϕ is any **test function** belonging to $\mathcal{F}_T$.

The condition expressed by the eq. 3.26 is manifestly weaker, i.e. less demanding, than the corresponding strong form, eq. 3.25. Indeed, any function solving 3.25 also solves 3.26 while the reverse is not true.

In fact by letting $\tilde{f} \equiv (\partial_t - \mathcal{L}) f$, the eq. 3.26 only implies orthogonality in $\mathcal{F}_T$ between ϕ and $\tilde{f}$.

$$(\phi, \tilde{f}) = 0 \tag{3.27}$$

This is much weaker than the strong solution requirement $\tilde{f} = 0$.

To obtain a finite set of evolutions equations, the infinite-dimensional functional space $\mathcal{F}_T$ is replaced by a finite-dimensional counterpart $\mathcal{F}_T^{(N)} = \{\phi_1, \phi_2, ...\phi_N\}$, spanned by N test functions. By plugging the expansion eq. 3.24 into eq. 3.26, and identifying ϕ in turn with $\phi_1, \phi_2, ...\phi_N$, we obtain the following system of ODE's:

$$\sum_{\mu=1}^{N} A_{\nu\mu} \dot{f}_\mu = \sum_{\mu=1}^{N} B_{\nu\mu} f_\mu, \; \nu = 1, N \tag{3.28}$$

where:

$$\begin{aligned} A_{\nu\mu} &= (\phi_\nu, \psi_\mu) \\ B_{\nu\mu} &= (\phi_\nu, \mathcal{L}\psi_\mu) \end{aligned} \tag{3.29}$$

are the "mass" and "force" matrices respectively. Due to the local nature of the operators, as well as of the basis functions $\{\psi_\nu\}$, both the mass and force matrices are generally very sparse, the degree of sparsity depending on the specific order of the polynomial representing ψ_ν and ϕ_μ .

For instance, in the simple case of piece-wise linear functions $e_\nu(x)$ it is readily checked that these matrices take a tridiagonal form because each element overlaps only with the left and right nearest neighbors.

It's worth noting that, at variance with both FD and FV, the mass matrix is generally **non-diagonal**, because of the element overlapping. This means that, although A and B are both sparse, the system described by eq. 3.28 is non-local in the sense that the rate of change of a single unknown $\dot{f}_\nu$ depends upon all of the other values via the mediation of the full matrix $A^{-1}B$.

This is sometimes regarded as a drawback of FE with respect to FD and FV. We shall return later on this point when discussing the time-integration procedures.

For the time being, we would like to point out a few merits of FE. First, its extreme generality as a procedure to derive finite-difference schemes for PDE's. This generality stems from the strong ties of the FE to the foundations of approximation theory in functional spaces.

Second, the elegant incorporation of Dirichlet and Von Neumann boundary conditions within the weak formulation underlying the FE via integration by parts.

Third, sistematicity; once the basic choices in terms of the type of elements to be used and boundary conditions is made, the route towards the construction of the finite-difference equations is absolutely well-defined with no degree of arbitrariness left to the analyst.

This systematicity is well highlighted by the process of matrix assembly, i.e. the formation of matrices A and B.

Here, one typically concentrates on a single element Ω_e and forms a **local** matrix $A_{lm}^{(e)}, l, m = 1, NN$, where NN is the number of nodes forming the element. Take, for instance, triangulated domain such as the one depicted in Fig. 3.9 and let's assume that linear (pyramid) basis functions $\Pi_\nu(x, y)$ are used.

For each triangle we can form $9 = 3 \times 3$ non-zero contributions (we refer here to the mass matrix for simplicity)

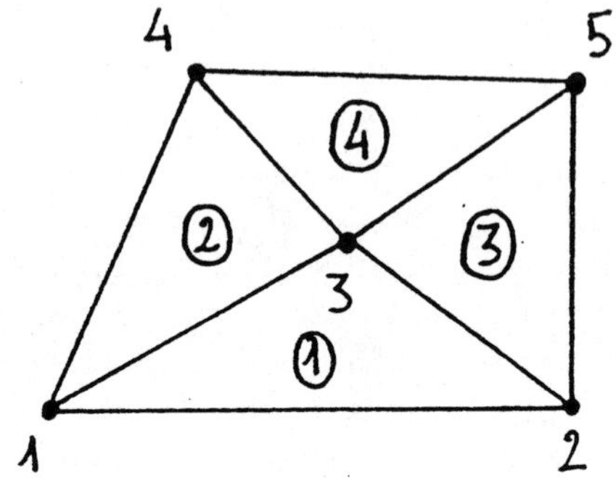

Figure 3.9: A four element triangulation

$$A_{lm}^{(e)} = (\Pi_l^{(e)}, \Pi_m^{(e)}) \tag{3.30}$$

where $l, m = 1, 3$ are local indices running along the vertices of the triangle. Assuming a counterclockwise convention for local indices, for the element Ω_3, we have the following correspondence (Fig. 3.10):

As a result, the local "submatrix" $A_{lm}^{(3)}$ contributes to the global matrix $A_{\nu\mu}$, as follows

$$A_{11}^{(3)} \rightarrow A_{22}, \quad A_{12}^{(3)} \rightarrow A_{25}, \quad A_{13}^{(3)} \rightarrow A_{23} \tag{3.31}$$

The final global matrix can therefore be formed by summing over all the element contributions:

$$A_{\nu\mu} = \sum_{l=1}^{NE} A_{lm}^{(e)} \tag{3.32}$$

This discussion highlights a fundamental aspect of unstructured schemes: the need of an (indirect) addressing array mapping local addresses onto global ones.

$$\begin{array}{lll} \nu = I(l; e) & e = 1, NE & \nu, \mu = 1, NN \\ \mu = I(m; e) & l, m = 1, LN & \end{array} \tag{3.33}$$

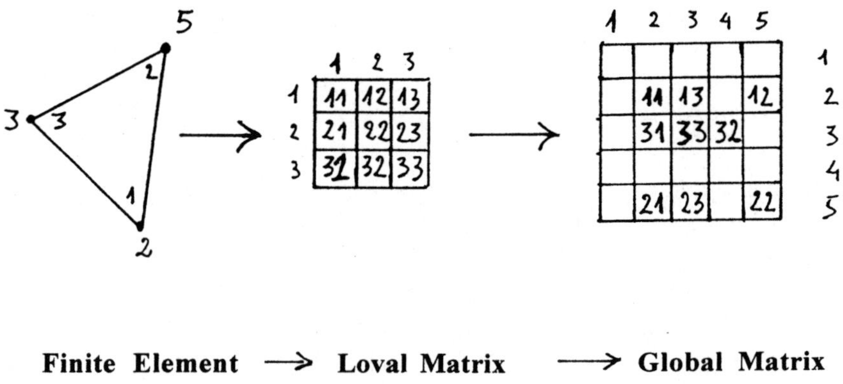

Figure 3.10: The process of matrix assembly in a finite-element calculation

where LN is the number of local nodes forming a given element.

Again, we see a blame and a blessing of the FEM. On the one side, the addressing array is all what we need to account for the lack of structure of the computational grid. Conceptually this is very elegant; once the array I has been agreed upon, one can proceed "blindly" element-by-element without even caring how many neighbors a given node interacts with: the procedure will automatically generate the right answer. On the other hand, from the practical point of view, the need of consulting each time the addressing array to know "who is who in the mesh", can severely tax the efficiency of the scheme whenever the data access pattern becomes too irregular. This puts a strategic value on optimal **mesh numbering strategies** geared to minimizing the matrix bandwidth, i.e. the maximal logical distance $|\nu - \mu|$ between nodes sitting on physically neighboring positions.

As we have seen, the assembly task consists of two logical steps:

A1 - Form local contributions $A_{lm}^{(e)}$

A2 - Scatter them onto the appropriate global addresses $A_{\nu\mu}$

It should be noted that $A1$ usually requires numerical quadrature to evaluate the integrals $A_{lm}^{(e)}$.

Typically, in flow situations, we have three type of contributions [42]:

"**mass-like**" $A_{lm}^{(e)} = (\phi_l, \psi_m) = \int_{\Omega_e} \phi_l(\vec{x})\psi_m(\vec{x})d\vec{x}$

"**force-convection-like**" $C_{lm}^{(e)} = (\phi_l, \psi_m')$ where the prime denotes space derivative

"**force-diffusion-like**" $D_{lm}^{(e)} = (\phi_l, \psi_m'') = -(\phi_l', \psi_m')$

where prime denote space derivative and integration by parts has been used to remove higher order derivatives.

Even though these integrals could often be computed analytically, it's a common practice to evaluate them by means of numerical quadrature, to preserve generality and ensure higher flexibility of the computer code.

The order of the quadrature is chosen such as to preserve the accuracy of the solution provided by the order of the elements. In general, for well-posed problems entailing derivatives up to order d, by using polynomials of degree p, the error should scale as $E_h \simeq O(h^{p+1-d})$ where h is a typical linear size of the element and the error is evaluated in L_2 norm: $E_h = (\int(f - f_h)^2 d\vec{x})^{1/2}$, f_h being the approximate solution.

3.2.4 Spectral methods

The three techniques described so far, FD, FV, FE, can be classified as **local** (whence their common prefix "finite") in that the unknowns one solves for represent the shape of the unknown function in a localized neighborhood of a single spatial location (node).

Even when the mathematical procedure behind the formulation of the method relies upon a global principle, such as the weak-formulation, or on an integral construct, like Gauss-theorem, locality is carefully preserved by using basis functions with **local** support.

Keeping tight to this locality principle has several advantages in terms of computational efficiency, and also provides the possibility to handle complex geometrical boundary conditions.

If, however, geometry is not a major issue, then other representations may be better suited to capture the flow dynamics without necessarily sticking to the locality principle in configuration space.

To discuss this point in some more detail let us represent the unknown solution in terms of an expansion over a set of approximating functions $\{\phi_n(x)\}$ with the property of taking non vanishing values all over the computational domain:

$$f(\vec{x},t) = \sum_{n}^{N} f_n(t)\phi_n(x) \tag{3.34}$$

Although this expansion is formally identical to eq. 3.20, its physical interpretation is completely different. Since $\phi_n(x)$ are global functions, i.e. non-zero all over the domain Ω, the amplitudes $f_n(t)$ represent now a **global** property of the unknown function. Typically total mass, energy, or, as we shall see, the energy contained at a given scale of motion.

Notwithstanding with their global nature, the functions $\{\phi_n(x)\}$ can generate very efficient numerical schemes provided three conditions are met:

i - Boundary conditions are automatically satisfied by all the ϕ_n's.

ii - The ϕ_n's form an orthonormal system, i.e. $(\phi_{n'}\phi_m) = \delta_{nm}$

iii - Differential operators are represented by sparse matrices.

In view of ii), we have:

$$f_n(t) = (f_x, \phi_n) \equiv \int f(\vec{x},t)\phi_n(\vec{x})d\vec{x} \tag{3.35}$$

where f_n is the n-th "moment" of the function $f(x)$, whence the definition of "moment-method" sometimes used for this kind of techniques.

Perhaps, the most popular moment method for CFD is the **spectral method** (SM). SM is based on a discrete Fourier representation of the velocity field [43] [34]:

$$\vec{u}(\vec{x},t) = \sum_{n=0}^{N} \vec{u}_n(t)e^{-i\vec{k}_n\cdot\vec{x}} \tag{3.36}$$

As a result, the method is targeted to homogeneous flows with periodic boundary conditions. By imposing $\vec{u} = 0$ on the boundary of a cubic box of size L, one singles out the following discrete spectrum of wave numbers:

$$\vec{k}_n = \frac{2\pi}{L}\vec{n} = \frac{2\pi}{L}(n_x, n_y, n_z) \tag{3.37}$$

where $n_x, n_y, n_z = \pm 1 .... \pm N/2$ label increasingly higher order modes, i.e. smaller wavelengths: $l_n = (2\pi)/k_n = L/n$.

In view of the orthogonality of the set $\{\phi_n \equiv e^{-i\vec{k}_n \cdot \vec{x}}\}$, one can readily check that the quantity:

$$E_n = \frac{1}{2}|u_n|^2 = \frac{1}{2}u_n^* u_n \tag{3.38}$$

represents the total energy carried by the flow at scale $l_n = L/n$.
The most immediate advantage of working in Fourier space is two-fold:

- Space derivatives are represented by algebraic diagonal operators: $\partial/\partial\vec{x} \leftrightarrow i\,\vec{k_n}\delta_{nm}$
- This representation is "exact" in the sense that the replacement $\partial/\partial\vec{x} = i\vec{k_n}$ is exponentially accurate in the limit of N going to infinity.

The other side of the medal is that **non-linear** convective terms, which are local in configuration space, become highly non-local in $\vec{k}$ space due to the well-known effect of non-linear harmonics generation. For the case of the Navier-Stokes equations, by applying the convolution theorem to the non-linear advection operator we obtain:

$$\vec{u} \cdot \frac{\partial \vec{u}}{\partial \vec{x}} = -i \sum_{k_p + k_q = k_n} \vec{u}_p \cdot \vec{k}_q \vec{u}_q \equiv C(\vec{u}, \vec{\nabla}\vec{u}) \tag{3.39}$$

This means that any mode interacts with **any other** in a completely global fashion. This gives rise to a quadratic $O(N^2)$ computational complexity which makes SM fairly unpractical.

The way out of this bottleneck is the **Pseudo-spectral** method [44]. This method consists of replacing the convolution

$$C(\vec{u}, \vec{\nabla}\vec{u})$$

with the Fourier transform of the products of the anti-transform of $\vec{u}$ and $\vec{\nabla}\vec{u}$ separately. Symbolically:

$$C(\vec{u}, \vec{\nabla}\vec{u}) = F[\vec{u} \cdot \vec{\nabla}\vec{u}] \simeq F\{F^{-1}[\tilde{u}_k \cdot F^{-1}[i\vec{k}\tilde{u}_k]\} \tag{3.40}$$

where $\tilde{u}_k$ denotes the Fourier-transform of $\vec{u}$: $\tilde{u}_k \equiv F[\vec{u}], \vec{u} = F^{-1}[\tilde{u}_k]$.
This turns around the $O(N^2)$ problem at the price of three additional transforms, one direct and two inverse.

Since discrete Fourier transforms can be executed quite efficiently on digital computers via the celebrated FFT (Fast-Fourier Transform) algorithm in $N \log N$ operations, the eq. 3.40 is a definite win. Note that the equivalence between $C(\vec{u}, \vec{\nabla}\vec{u})$ and the r.h.s. of eq. 3.40 is only an approximate one in a finite domain ($L < \infty$) because of aliasing effects, i.e. unresolved frequencies which contaminate the resolved ones ($-N/2 \leq p, q \leq N/2$, but $|p+q| > N/2$).

De-aliasing is therefore required in the form of a sharp truncation of the modes beyond a given threshold: $|u_n| = 0$ for $|n| \geq |n_{max}| = N/\sqrt{2}$.

Pseudo-spectral methods have been the workhorse for direct simulation of fluid turbulence for more than two decades now.

They have been extended to simple bounded flow geometries by use of Tsebitchev polynomials [45] and more recently, they have been merged with the finite element method to combine the best features of both techniques within the so called Spectral Element Method [46], [47].

3.3 Time-discretization and non-linearity

The methods described so far can be broadly categorized as alternative ways of taking a PDE into a convenient set of ordinary differential equations (ODE). In order to make this ODE's amenable to computer manipulation, they have to be turned into suitable ADE's (algebraic difference equations): this task is accomplished by discretizing the time variables ("time marching schemes") [48].

Let us begin with grid-methods: FD, FV, FE. The ODE's associated with these methods can be cast into the following general form:

$$\sum_{m=1}^{NH} A_{lm} \dot{f}_m = \sum_{m=1}^{NH} B_{lm}(f_k) f_m \tag{3.41}$$

where NH is the number of neighbors of f_l.

An important class of time-marching schemes is represented by the following one-step expression:

$$\begin{aligned} \dot{f} &\to (\hat{f} - f)/\Delta t \\ f &\to \tilde{f} \equiv \hat{\theta} f + \theta \hat{f} \end{aligned} \tag{3.42}$$

where caret indicates the "future" instant t_{n+1} and no-caret the actual one, t_n. Here θ is an interpolation parameter ranging from 0 to 1, and $\hat{\theta} = 1 - \theta$, which controls the degree of **causality** of the time-marching scheme.

By inserting the relations 3.42 into the eq. 3.41 we obtain the following system of non-linear ADE's:

$$\mathcal{A}(\tilde{f})\hat{f} = \mathcal{B}(\tilde{f})f \tag{3.43}$$

where

$$\begin{aligned} \mathcal{A} &= (A - \tilde{B}\Delta t\theta) \\ \mathcal{B} &= (A + \tilde{B}\Delta t\hat{\theta}) \end{aligned} \tag{3.44}$$

and subscripts have been omitted for the sake of notational simplicity. Here $\tilde{B} \equiv B(\tilde{f})$.

The formal solution of the eqs. 3.43 and 3.44 is given by

$$\hat{f} = \mathcal{A}^{-1}\mathcal{B}f \equiv Tf \tag{3.45}$$

where T is the "transition matrix" taking the state of the system at time t into the corresponding state at time $t + \Delta t$. The transition matrix plays a fundamental role in the numerical solution of the Navier-Stokes equation and consequently a few comments on its properties are in order.

A first observation is that, in spite of the local nature of the ODE's it derives from, the transition matrix is generally dense thus expressing a **global** dependence: the actual state of a given unknown at a certain spatial location is generally dependent on the previous state of **all** other unknowns.

Formally, this is due to the fact that, although the matrices $\mathcal{A}$ and $\mathcal{B}$ are sparse, the inverse matrix $\mathcal{A}^{-1}$ is not.

The origin of this non-locality is the presence of non-causal links represented by the matrix $\mathcal{A}$, which couples simultaneous values of the unknowns at spatially distinct locations, thus introducing "de-facto" an unphysical action-at-distance effect.

There is a wide class of algorithms for which space-time locality can however be fully recovered. This class is identified by the following conditions

$$\begin{aligned} &\theta \to 0 \quad (Explicit) \\ &A \to I \quad (Diagonal) \end{aligned} \tag{3.46}$$

The first condition identifies **explicit** schemes, i.e. those for which future values at a given site only depend on the present values of a restricted neighborhood (domain of influence). The extent of this neighborhood is controlled by the bandwidth of the transition matrix.

The second condition ensures that the bandwidth of the transition matrix is just the same as the one of B; in other words, the domain of influence matches with the original stencil of the differential operator.

Explicit *and* diagonal schemes are **matrix-free**: they don't require the solution of any algebraic system to obtain the future state as a function of the present one.

This implies considerable simplifications in the computer codes besides obvious memory economies.

Explicit FD and FV fall within this class, while FE generally doesn't because, due to element overlapping, A is sparse but not strictly diagonal. This is often regarded to as a definite downside of FE, although in some instances locality can be artificially restored by the "mass lumping" trick: $A \rightarrow I$, i.e. by barely replacing A with the unit matrix. Mass lumping may however lead to instabilities and lack of convergence for convection-dominated flows.

The parameter θ measures the degree of impliciteness of the time-marching scheme, $\theta = 1$ being associated with fully impliciteness, and can be adjusted to achieve 2nd order accuracy in time ($\theta = 1/2$, known as Cranc-Nicolson scheme).

Explicit schemes are physically sound and simple, but these virtues are bought at the expense of strict limitations on the maximum allowed size of the time-step Δt. This is subject to the CFL (Courant-Friedricks-Levy) conditions:

$$u\Delta t \leq \Delta x; \quad \nu\Delta t \leq \Delta x^2 \tag{3.47}$$

which cannot be violated without incurring in disruptive numerical instabilities, i.e. numerical blow-up due to the fact that the transition matrix is a dilatation ($||T|| > 1$) rather than a contraction with the result that numerical errors grow instead of decaying.

Again, the CFL conditions are nothing but a mirror of causality: the physical information cannot travel faster than the highest speed allowed by the discrete mesh ($\Delta x/\Delta t$).

Implicit schemes, doing away with locality, are freed from CFL limitations and consequently they can march in much larger steps without suffering any numerical instability (fig. 3.11, 3.12) [54].

Obviously, this requires much more work per time-step, since at each time-step a set of ADE's has to be solved. In addition it is clear that no

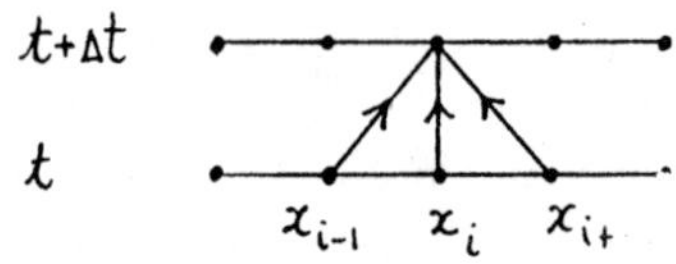

Fully Explicit

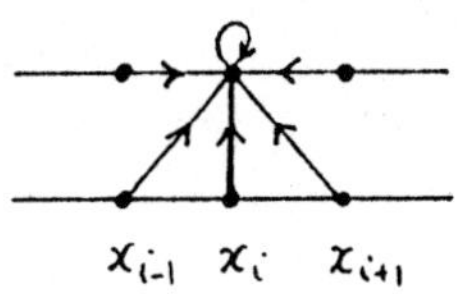

Fully Implicit

Figure 3.11: The space-time connectivity of explicit and implicit method respectively. Lines on the same time level break causality

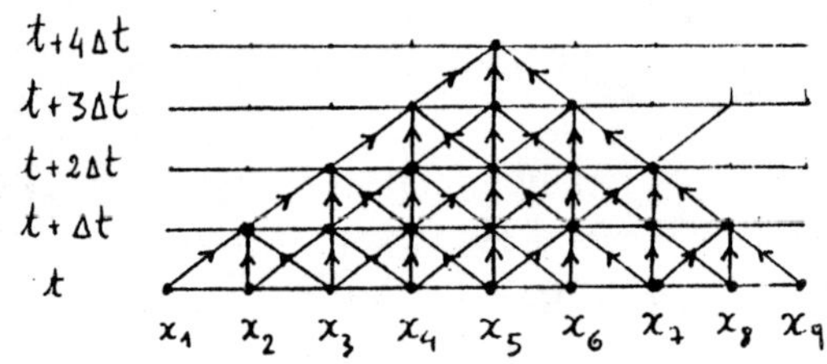

Explicit; Δt_{exp}=Δt

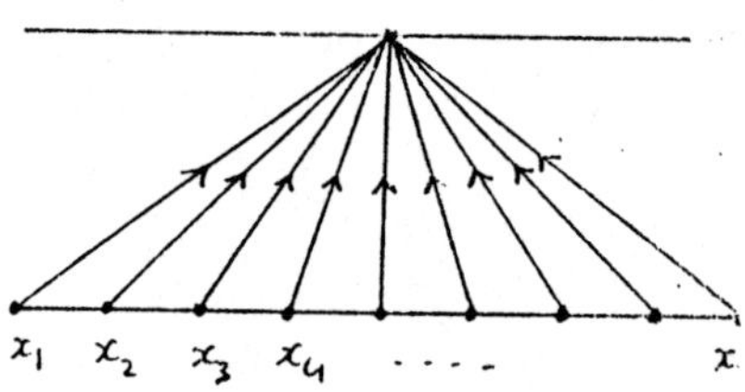

Implicit; Δt_{imp}=4Δt

Figure 3.12: Time marching with explicit and implicit schemes

physics at scales below Δt can be explored because all frequencies above $2\pi/\Delta t$ are automatically filtered out.

For slow flows, or for situations where only steady-state conditions are sought for, the latter disadvantages are often more than compensated by the gain in time-step size, and consequently implicit methods are the method of choice. On the other hand, for highly active flows, such as those which develop dynamic instabilities taking from laminar to turbulent regimes, it is clear that explicit methods are to be preferred.

Under such circumstances, high temporal accuracy may be desirable which points to multistep Runge-Kupta integrators.

3.3.1 Non-linearity

Within grid methods, the Navier-Stokes non-linearities remain localized in space and can be effectively handled by means of recursive iterations. A typical iteration cycle may look as follows:

$$\begin{aligned} \mathcal{A}(\hat{f}^k)\hat{f}^{(k+1)} &= \mathcal{B}(\hat{f}^k)f \equiv g^{(k)} \\ f^{(0)} = f; \qquad & k = 1, 2, \ldots.., NIT \end{aligned} \tag{3.48}$$

where $\hat{f}^{(k+1)}$ is the actual unknown at $(k+1) - th$ cycle.
The iteration loop is closed once the stopping criterion, of the form:

$$\epsilon^{(k+1)} \equiv \left| \frac{\hat{f}^{(k+1)} - \hat{f}^{(k)}}{\hat{f}^{(k)}} \right| < \epsilon \tag{3.49}$$

is fulfilled for some $k < NITMAX$, where $NITMAX$ is the maximum allowed number of outer iterations. Here $||.||$ denotes some suitable norm, and ϵ is a preselected error tolerance.

The eq. 3.48 represents a sequence of **linear(ized)** ADE's which are to be solved at each time-step until the convergence criterion is matched.

The iterative loop dealing with nonlinearity is commonly referred to as to *outer* iteration loop.

3.4 Solving linear algebraic equations

The solution of the linear systems of ADE's , eq. 3.48, is often the most intensive computational kernel of CFD codes. Here, the solution methods fall within two broad classes: **direct** and **iterative** solvers [49] [50].

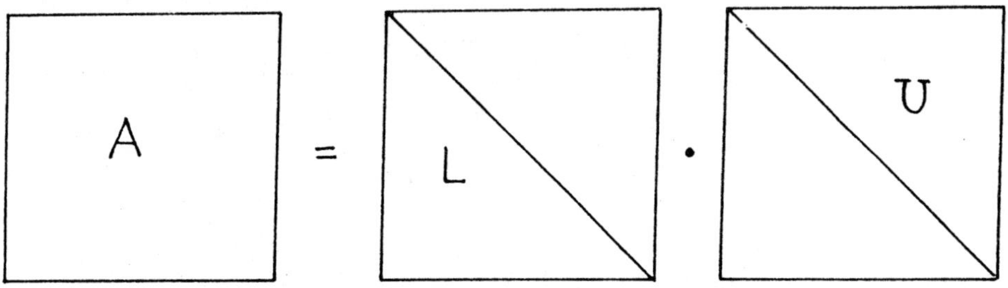

Figure 3.13: The LU-decomposition of matrix A

3.4.1 Direct methods

Direct methods are based on Gauss elimination techniques aimed at decomposing the matrix A (we drop the notation $\mathcal{A}$ which is no longer needed at this stage) in a form suitable to straightforward solution via direct substitution. A typical example is the LU decomposition, $A = LU$, L and U being down-left and upper-right matrices respectively (fig. 3.13).

By letting $h = Uf$ and $Lh = g$, the original system $Af = g$ is solved as a two-step sequence of forward-backward sweeps:

- $h = L^{-1}g$ (forward sweep)
- $f = U^{-1}h$ (backward sweep)

Direct methods are very safe and robust (eventually round-off errors should be carefully monitored) but suffer of the serious draw-back of "fill-in", i.e. dynamic generation of non-zero entries in the course of Gauss substitution $a_{ij} \to a_{ij} - \sum_k (a_{ik}a_{kj}/a_{kk})$.

This means that storage must be reserved also for zero entries in the matrix A.

The result is that direct methods scale like

- Storage $= S \simeq NB_W$

- Computer work = $W \simeq N B_W^2$

where B_W is the matrix **bandwidth**, the maximal logical distance between interacting elements. The point is that, even if the matrix A is sparse, i.e. few non-zero entries, the bandwidth B_W can be of the order of L^{D-1} where L is a linear size of the domain and D is the number of physical dimensions.

This yields:

$$S \simeq N^{2-\frac{1}{D}}, \qquad W \simeq N^{3-\frac{2}{D}} \tag{3.50}$$

This unfavourable scaling makes direct methods very unpractical for large problems in more than two-dimensions.

The "fill-in" problem can be alleviated by using smart numbering schemes and memorisation formats.

However, the most radical move to do away with fill-in altogether is to use iterative methods.

3.4.2 Iterative methods

Iterative methods come in two basic flavors [51];

- Relaxation schemes
- Gradient schemes

Both these classes of methods work on the principle of finding a sequence of **approximate** solutions $\{f^{(0)}, f^{(1)}, ..., f^{(k)}\}$ of the linear algebraic problem:

$$Af = g \tag{3.51}$$

converging to the exact solution, f^*, in the least possible number of steps.

As usual, convergence is intended in the sense of getting "sufficiently close to":

$$\epsilon^{(k)} = ||\frac{f^{(k)} - f^*}{f^*}|| < \epsilon \tag{3.52}$$

A wide choice of methods is available to this end, all of which share the common feature of relying upon **sparse** matrix manipulations throughout the sequence of steps leading to the final solution.

Relaxation schemes are quite conveniently formulated in terms of "fake-time" evolutionary problems:

$$\frac{df}{d\tau} = g - Af \tag{3.53}$$

where τ is the fake-time parametrizing the family of approximants $f^{(k)} = f(\tau_k); \tau_k = k$.

Various iterative schemes can be generated depending upon the sort of discretization adopted to march the eq. 3.53 in fake-time.

Popular choices of the two-steps splitting scheme: $Af \simeq A_1 f^{(k)} + A_2 f^{(k+1)}$, are:

- **Picard's scheme**

$$Af \simeq Af^{(k)} \qquad (A_1 = A; A_2 = 0) \tag{3.54}$$

- **Jacobi**

$$Af \simeq Df^{(k+1)} + (L+R)f^{(k)} \qquad (A_1 = D; A_2 = L+R) \tag{3.55}$$

- **Gauss-Seidel**

$$Af \simeq (L+D)f^{(k+1)} + Rf^{(k)} \qquad (A_1 = L+D; A_2 = R) \tag{3.56}$$

where $A = L + D + R$, with L upper left, R upper right and D diagonal.

By replacing $df/d\tau$ with $(f^{k+1} - f^k)$, all of these schemes take the general form:

$$(I - A_1)f^{k+1} = (I - A_2)f^{(k)} + g \tag{3.57}$$

where $(I - A_1)$ is at most lower-triangular and consequently does not require any linear algebra solver. Very often, to speed-up convergence, a relaxation step is applied

$$f^{k+1} \leftarrow (1-\omega)f^{(k)} + \omega f^{k+1} \tag{3.58}$$

where ω is a relaxation parameter between 0 and 1 (under-relaxation) or 1 and 2 (over-relaxation).

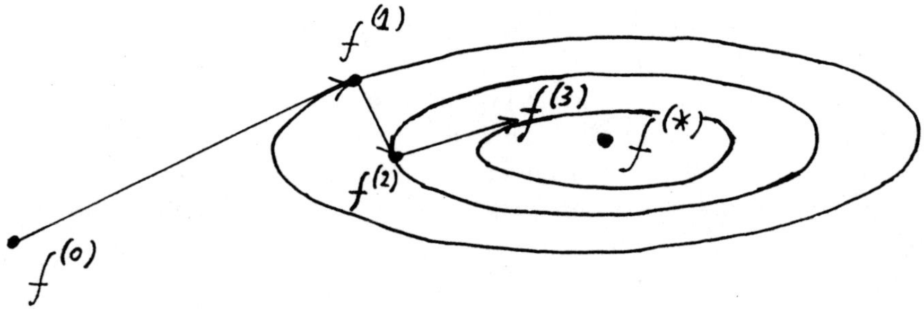

Figure 3.14: A search trajectory leading to the bottom of the "valley" defined by the dissipation potential

Gradient methods are based on the idea of seeking the minimum of the dissipation potential

$$\Phi = (f, g - Af) \tag{3.59}$$

where $(\cdot,\cdot)$ denotes scalar product in $\mathcal{R}^N$. By definition, this minimum corresponds to the exact solution.

By dotting the eq 3.59 with f we obtain

$$\frac{dK}{dt} = \Phi \tag{3.60}$$

where $K = 1/2(f,f)$ is the "kinetic-energy" of the system, to be seen as a material point moving in a N-dimensional space under the effect of the "force" g and slowed down by the friction term Af.

The solution of the algebraic system, eq. 3.60, corresponds to the equilibrium, rest position, $f = f^*$. The idea is then to find the shortest (in terms of number of steps) path taking the initial guess $f^{(0)}$ to within a small neighborhood of f^*.

Iterative schemes in this class fall within the following general form:

$$f^{(k+1)} = f^{(k)} + \lambda_k s^{(k)} \tag{3.61}$$

where $s^{(k)}$ is the search direction at step k, and λ_k is a scalar coordinate whose value is obtained by minimizing Φ along direction $s^{(k)}$.

Among the most popular methods, we list the following:

Steepest Descent (SD)

Here, the search is carried out along the direction of maximal change of the dissipation function Φ, i.e. the direction identified by $\text{grad}\Phi = Af - g \equiv r$ (residual): $s^{(k)} = r^{(k)}$; $(r^{(k+1)} \cdot r^{(k)} = 0)$.

The SD method progressively finds minima in subspaces of $\mathcal{R}^N$ by starting from a one-dimensional space and progressively expanding it to a N-dimensional one. When the dimensions of the search space matches the dimensionality of the problem, the solution is found (in principle) in N steps.

Conjugate Gradient (CG) [52]

The SD method can be extremely slow if the ratio of the max to min eigenvalues of A (called **condition number**) is large, i.e. if the valley of the function Φ is very long and narrow

This problem is considerably alleviated by allowing a certain "memory" within the search process; more specifically the new search direction $s^{(k+1)}$ is expressed as a linear combination of the **actual** residual $r^{(k+1)}$ and the previous one, $s^{(k)}$:

$$s^{(k+1)} = r^{(k+1)} + \beta s^{(k)} \tag{3.62}$$

The coefficients (λ, β) can then be found by locating the minimum of $\Phi(\lambda, \beta)$ in the **plane** (s^k, s^{k+1}). It is readily proved that this leads to perform the search along the conjugate direction s^{k+1}, defined by the A-orthogonality condition $(s^{k+1}, As^k) = 0$ rather than along a sequence of orthogonal directions like in SD.

The CG algorithm can be synthetized as follows:

1. Start from a first guess $f^{(0)}$
2. Compute $s^{(0)} = r^{(0)} = Af^{(0)} - g$
3. $\lambda = (r, r)/(s, As)$
4. $\hat{f} = f + \lambda s$
5. $\hat{r} = r - \lambda As$
6. $\beta = (\hat{r}, \hat{r})/(r, r)$

7. $\hat{s} = \hat{r} + \beta s$

8. go to 2 (if not converged)

where caret denotes the $(k+1)$-th cycle.

Convergence in N steps would be guaranteed in the absence of round-off errors. The latter often result in a significant deterioration of the convergence, especially if the matrix A is ill-conditioned.

This calls for the need of up-fronting the CG method with appropriate **preconditioning** techniques, aimed at transforming the linear system: $Af = g$ into an equivalent but better conditioned one: $\tilde{A}\tilde{f} = \tilde{g}$, where $\tilde{A}$ relates to A by a similarity transformation.

The idea is to have $\tilde{A}$ as close as possible to the unit matrix I (perfectly well conditioned problem).

One popular class of techniques to get close to this target is presented by **"incomplete" decomposition** methods. Within these techniques, one looks for an approximate decomposition of the matrix A in the LDU form:

$$A = LDU \simeq L_A D_A U_A \tag{3.63}$$

where the subscripted matrices $L_A D_A U_A$ (approximate) differ from the exact ones in that they are forced to retain the same sparsity structure of the matrix A.

The original system $Af = g$ can consequently be re-expressed as

$$D_A^{-1} L_A^{-1} A U_A^{-1} f = D_A^{-1} L_A^{-1} g \equiv \tilde{g} \tag{3.64}$$

where $U_A f = \tilde{f}, \tilde{A} \simeq I$ so long as $L_A \simeq L$ and $D_A \simeq D, U_A \simeq U$. The original problem therefore splits into three steps:

- Compute transformed r.h.s. $\tilde{g} = D_A^{-1} L_A^{-1} g$
- Solve iteratively for the transformed unknown $\tilde{f} : \tilde{A}\tilde{f} = \tilde{g}$
- Back to the original unknown $f : U_A f = \tilde{f}$

The success of gradient-like iterative methods hinges on the matrix $\tilde{A}$ being positive definite and well-conditioned. These conditions are commonly met for diffusion-dominated flows (low Reynolds, parabolic problems) with smooth diffusivities.

For high-Reynolds flow the matrix A is no longer symmetric and positive definite. To face these problems, special variants of the CG scheme have been developed which are documented in the literature [53].

Further details on this fast-moving topic go far beyond the scope of this book.

Here we shall conclude by pinponting the fact that preconditioned schemes are "de-facto" a blend of iterative and direct methods. In fact, in order to perform the incomplete decomposition eq. 3.63, a direct solver is required.

Gradient-methods, and the related variants, have gained an increasing popularity within the CFD community dealing with complex geometries. This is mainly due to the great generality of the method, which does not rely upon any simplifying assumption on the nature of the flow. For strongly coupled, high-Reynolds flows, any iteration scheme, no matter how robust, may however fail to converge. Under these circumstance, direct solvers are mandatory.

As a matter of fact, one may easily predict that in the near future direct and iterative solvers will be combined together in different regions of the flow to produce optimal hybrid solution strategies.

Chapter 4

Numerical solution of fluid flow equations: particle methods

4.1 Particle methods

The common objective of grid methods is to turn a system of partial differential equations (PDE's) into a corresponding system of algebraic difference equations (ADE's). This task is accomplished in two steps: first PDE's are converted into ordinary differential equations (ODE's) via discretization of the spatial variables, then ODE's are converted into ADE's via time discretization:

From this broad perspective, all the grid methods discussed so far can be seen just as alternative ways to accomplish step 1 by means of different numerical discretization devices.

$$
\begin{array}{ccccc}
PDE's & \longrightarrow & ODE's & \longrightarrow & ADE's \\
 & \uparrow & & \uparrow & \\
 & \textit{space} & & \textit{time} & \\
 & \textit{discretization} & & \textit{discretization} &
\end{array}
$$

Figure 4.1: The logical steps of the discretization procedure taking a PDE into a set of ADE's to be solved on a computer.

The distinctive feature of Particle Methods (PM) is that step 1 is no longer needed because the physical model is formulated in terms of ODE's right at the outset. These ODE's are nothing but the equations of motion of **discrete** (as opposed to "discretized") entities that we generally classify as "particles" [55]. The key observation is that computational particles need **not** be identified with real-world particles. Actually they **should** not, for this would entail a degree of detail leading to an astronomical amount of information which is both unmanageable and highly redundant to the purpose of describing the dynamics of a fluid system. As a result, the key to success of a smart PM scheme is to sort out the most convenient mapping between physical particles and computational ones.

The smartest methods are those who succeed in getting rid of all physical irrelevant details (the ones that are anyway eliminated from the physical picture while moving top-bottom along the BBGKY hierarchy) without loosing the crucial features which characterize fluid flows.

Let us proceed with a cursory view of the some representative examples in this breed, i.e.:

- Molecular Dynamics (MD)
- Lattice Gas Automata (LGCA) and Lattice Boltzmann (LB)
- Direct Simulation Montecarlo (DSMC)
- Vortex Methods (VM)
- Smoothed Particle Hydrodynamics (SPH)

4.1.1 Molecular dynamics

In the real world, fluid dynamics arises from the statistical average of the motion of a very large number ($O(10^{20}/cm^3)$) of molecules, obeying Newton's classical equation of motion

$$\begin{cases} \dot{\vec{x}}_p = \vec{v}_p \\ \dot{\vec{v}}_p = \frac{1}{m}(\sum_q \vec{F}_{pq} + \sum_{qr} \vec{F}_{pqr} + ...); \qquad p = 1, ..., N_P \end{cases} \tag{4.1}$$

These molecules stream freely in physical space and undergo collisions once they come sufficiently close together. Collisions can be extremely complex depending on the actual form of the interaction force $\vec{F}_p = \sum_q \vec{F}_{pq} + \sum_{qr} \vec{F}_{pqr} + ...$ as a function of the particle coordinates. Regardless of their complexity, these interactions are bound to conserve local mass, momentum and energy. This is the key ingredient to recover fluid behaviour at a macroscopic level.

MD is not targeted to Fluid Dynamics: instead it has been used for years to deduce macroscopic properties, such as viscosity, conductivity out of the microscopic interaction laws; in a word, to bridge the gap between the micro and the macroworld [56] [57] [58].

A new direction, which is surfacing with the advent of parallel computers, is that computers are becoming fast and powerful enough for MD modeling to extend all the way from the atomistic level up to the fluid regime.

Where does the atomistic "way" draw its appeal from? The main reason is that MD is a **grid-free** method: particle coordinates and speeds need not be attached to any geometrical mesh. This opens up two interesting opportunities.

- - Lenghty and difficult mesh generation procedures can bypassed. Boundary conditions can be easily implemented by knowing just the surface coordinates of the solid bodies without having to mesh the entire volume.
- - The method is intrinsically Lagrangian, i.e. the degrees of freedom (d.o.f.) are attached to particles rather than grid-points (Eulerian). This means that, in principle, the discrete d.o.f. should "flow" there where they are actually needed, in a sort of **naturally adaptive** fashion. This opens up a great potential for savings in computer storage and CPU.

From the computational point of view, the major problems to be overcome to take MD to the fluid scale are basically two: statistical noise and efficient representation of the interparticle interactions.

Statistical noise relates to the fact that the number of particles required to keep statistical fluctuations below an acceptable threshold can be unacceptably large even for small volumes of the fluid.

Efficient representation of the interparticle interactions tantamounts to work out "ad hoc" forms of the forces which must be simple enough to allow practical computations, without loosing at the same time, physical fidelity.

The practical issue, here, is "how many neighbors should a particle see in order for the whole ensemble of particles to behave like a fluid rather than like a 'wild bunch' of particles ?".

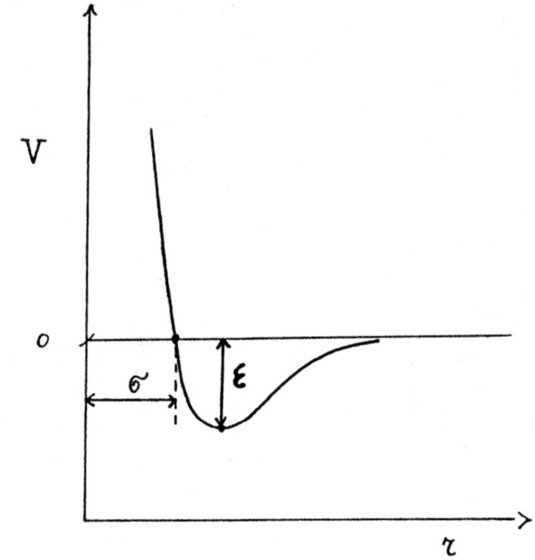

Figure 4.2: The Lennard-Jones potential as a function of the radial coordinate

One of the most popular choices to model interparticle collisions in MD simulations is represented by the truncated two-particle Lennard-Jones (soft-sphere) potential:

$$V_{LJ}(r) = \begin{cases} 4\epsilon[(\frac{r}{\sigma})^{-12} - (\frac{r}{\sigma})^{-6}]; & 0 \leq r \leq r_c = 2^{1/6}\sigma \\ 0 & \text{elsewhere} \end{cases} \tag{4.2}$$

This potential is short-ranged , i.e. particles lying more than a distance r_c apart simply do not interact.

This drastic simplification is mandatory to beat the N_P^2 complexity of the collision term, $\sum_q \vec{F}_{pq}, p = 1, N_P, q = p, N_P$, by turning it into a $N_P \cdot (r_c/L)^3 \simeq N_P$ complexity. Technically, this is implemented by means of linked lists, i.e. lists of interacting neighbors with no preassigned structure. The simulation system divided into N_c cells of linear size r_c. Within each cell, for any particle p belonging to the cell there is a pointer P to a list of neighbors $q_1, q_2, ..., q_{NH}$ which necessarily belong to either the same cell or to the neighboring cells (8 in two dimensions, 26 in three), since they must fall within a sphere or radius r_c centered about p.

The first attempt to simulate hydrodynamics from the atomic scale date of the mid 80's, with the work of Clementi, Rapaport, Mareschal and co-

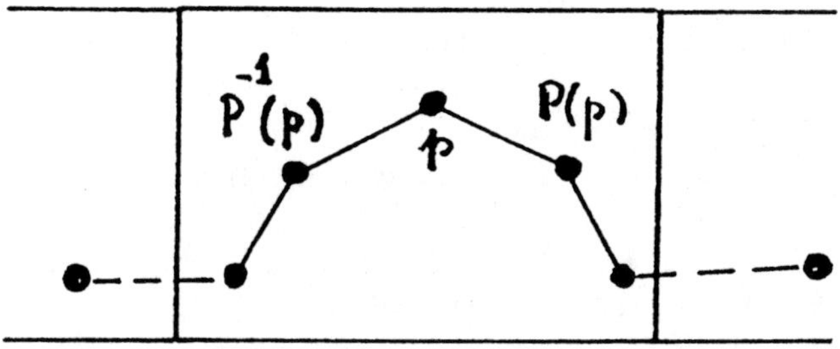

Figure 4.3: The neighbours of particle p: they fall within a sphere of radius r_c centered about particle p.

workers [59] [60]. These autors found that, quite remarkably, hydrodynamic behaviour sets in already at the scale of a few hundreds Angstroms !

MD simulations with $10^4 - 10^6$ particles have proved capable to reproduce several non-trivial fluid behaviours, such as the formation of vortices past bluff bodies and the formation of convection cells in Rayleigh-Bénard experiments (a fluid heated from below). This is very fascinating from the conceptual point of view, but still leaves a rough edge as long as computational efficiency is concerned. The main hurdle is the steep scaling of the Reynolds number with the number of particles needed in the simulation: $Re \simeq N_P^{1/3}$. This bounds MD to relatively low-Reynolds flows even on most powerful present-day computers ($N_P \simeq 10^6 - 10^7$) [61].

4.1.2 Direct Monte Carlo Simulation

Another important particle method, closely related to but also significantly distinct from Molecular Dynamics, is the Direct Simulation Montecarlo method (DSMC). Quoting its inventor, Graeme Bird, both MD and DSMC fit to the following definition:

A computer is used to provide direct physical simulation of fluid flow rather than a numerical solution of the equations that provide the mathe-

matical model of the flow [62], [63].

While sharing the same philosophy, and basically also the same chronological slot, (1958 MD and 1963 DSMC), the two methods differ in several important respects.

The most important one is the **probabilistic** nature of DSMC as opposed to the deterministic one in MD.

This is directly traced to the fact that MD works at the atomistic level of the BBGKY hierarchy, while DSMC deals with the kinetic level.

That's why, always quoting Bird:

"they're complementary rather than competitive when applications are considered".

Let us give a brief account on the basics of DSMC directing the reader keen of more detail to the specialized literature.

Like MD, and most particle methods for neutral fluids, DSMC splits the dynamics of the Boltzmann distribution function $f(x, v, t)$ in two separate steps: free-streaming and collisions.

Free-streaming proceeds exactly the same way as MD; particle coordinates and speeds are updated according to Newton's law. The probabilistic nature of DSMC shows up when it comes to the way that particle collision are dealt with.

To better focus on this point, let us rewrite the Boltzmann collision term

$$C(f, f) = \int [f(\vec{v})f(\vec{v}^*) - f(\vec{v}')f(\vec{v}'^*)] b\sigma(g) db d\vec{v}^* \tag{4.3}$$

As pointed out in Chapter II, this formula reflects the statement that the collision rate between any two molecules is proportional to the product of their one-particle distribution function (molecular chaos) times the volume swept over by their relative motion. Consider now a DSMC cell of size h and suppose that each computational molecule ("particle" thereafter) represents R real molecules (simply "molecules" hereafter). By definition:

$$np = nm/R \tag{4.4}$$

where np is the number of particles and nm is the number of molecules within the cell.

The probability P of a collision between two particles over the interval Δt is given by the fraction of volume swept by a particle over the cell volume, times the representative factor R:

$$P = Rg\sigma\Delta t/h^3 \tag{4.5}$$

The full set of collisions required to simulate the collision term is given by $R \times P \times np(np-1)/2$ single events.

This corresponds to selecting all possible interaction pairs and weight them with the corresponding probability. This 'naive' implementation is rather inefficient because, first, it exhibits quadratic complexity and second, because for many selected pairs P can take extremely small values so that they make little contribution to the collisionality.

Much better efficiency is obtained by selecting a proper subset of representative collisions and enhance their probability so as to preserve the same collisionality of the 'naive' procedure.

In practice, this means identifying the fraction of collisions with maximum probability P_{max} and restrict the collisional events to this subset with an enhanced weight $1/P_{max}$.

A remarkable feature is that if np is doubled by, say, halving the representativiness factor R, the computational work grows only **linearly** with the the number of particles.

Let us estimate the computational cost of DSMC.

As a reference frame, let us consider a three-dimensional flow extending M mean-free paths in each direction, and ask how many molecules are involved in such a flow. According to standard kinetic theory arguments, this number is given by

$$N_M = 4000\frac{n_l}{n^2}M^3 \tag{4.6}$$

where n is the gas number density and $n_l = 2.687\times 10^{25}$ is the Loschmidt number, i.e. the number of molecules per cubic meter under standard gas conditions.

This simple expression indicates that in order to simulate, say, 100 mean-free paths, we need about four billion molecules. This is roughly two orders of magnitude beyond current supercomputers capability.

Things get much better as density is increased. For instance, by raising n by a factor hundred, the number of molecules drops to a few hundred thousands, which is well within reach of present-day computers.

This is indeed the actual operating regime of MD simulations.

DSMC, however, is basically intended for **rarefied** flows, for which values of $n/n_l \simeq 10^{-6}$ are a common place. To simulate such flows, each particle must lump the information associated with about 10^{10} molecules!

A similar situation applies to macroscopic fluid flows extending over millions of mean-free paths (one free path in ordinary fluids is of the order of microns).

The conclusion is that for ordinary macroscopic flows, DSMC is far less efficient than "conventional" grid methods.

And yet, kinetic theory and particle methods still retain a keen interest for CFD, since they open up extra-perspectives in the definition of new and ever more powerful instances of the key concept of "computational molecule".

Lattice gas and lattice Boltzmann, to be described hereafter, provide two examples in this breed.

4.1.3 Lattice gas

The fact that as few as 1000 atoms already reproduce the motion of a fluid is an important conceptual breakthrough and one which sheds new hope on the possibility of the "atomistic-way" to CFD. Yet, from the practical point of view, MD is still way off less efficient than grid-based methods: flows which take MD hours and hours on supercomputers can be completed today on small PC's in a handful of minutes, if not seconds, with a good grid technique.

One possible way to improve the performance of MD is to further simplify the basic interaction between the computational atoms. The most radical move along this line was proposed in 1985, with the introduction of Lattice Gas Automata in a seminal paper by Frisch, Hasslacher and Pomeau [64].

These authors were able to prove that fluid dynamic behaviour can be reproduced by a system of fictious particles whose dynamics is simplified down to the point where it can be expressed by means of purely **logical** (boolean) operations.

The FHP lattice gas (LGA) automaton consists of a regular lattice with hexagonal symmetry. This lattice is populated by pseudoparticles hopping synchronously from site to site along the six possible directions of the grid (free streaming). Particles sitting on the **same** lattice site undergo discrete collisions, which allow them to exchange momentum.

A typical MOVE+COLLISION step is shown in the fig. 4.4

Both MOVE and COLLISION steps can be implemented in purely digital (boolean) form with no need of floating-point operations.

Suppose in fact to characterize the state of the automaton by a 6-bit word $n = (n_1, n_2, ..., n_6)$, where $n_i = 0, 1$ stand for absence/presence of a particle moving along the i-th direction.

With this representation, the MOVE step corresponds to scattering the bits into the proper locations over the six neighbors words (Fig. 4.5):

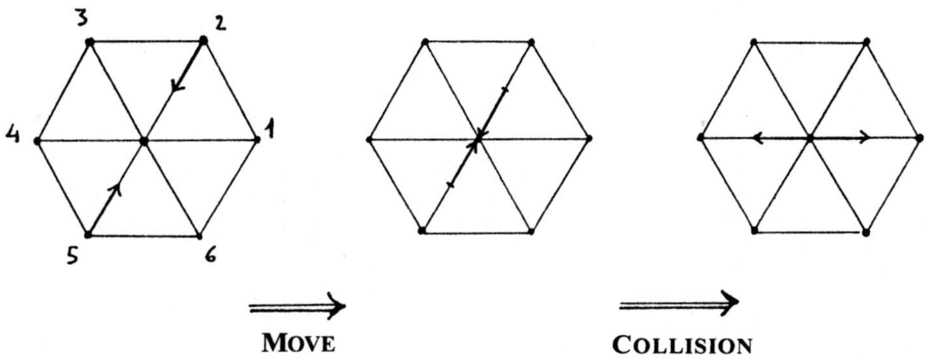

Figure 4.4: MOVE (free-streaming) and COLLISION in a FHP lattice gas

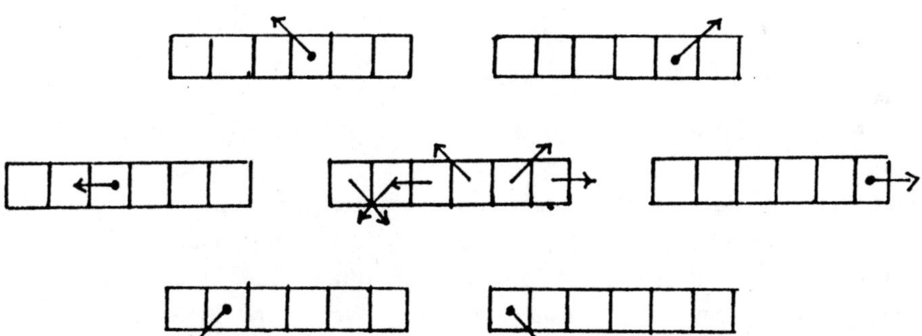

Figure 4.5: MOVE step in a FHP lattice gas: bits are numbered 1 to 6 counterclockwise from right to left

From the practical point of view, one must "extract" the bits from the source-word and "force" them to the proper location of the destination-word. These are clearly "exact" (no approximation due to the finite size of the words) operations which can be performed on any digital computer by means of elementary bit manipulation instructions.

A similar consideration applies to the COLLISION step, which corresponds to a **local** boolean transformation C of the string n_{in} (before the collision) into n_{out} (after collision).

$$n_{out} = C n_{in} \tag{4.7}$$

For instance, the head-on collision depicted in Fig. 4.5, corresponds to the following transformation:

$$010010 \longrightarrow 100100$$

The boolean map C sending n_{in} into n_{out} can be either constructed as an explicit combination of elementary boolean operations (AND, OR, NOT, XOR) or as a pre-constructed table look-up. In either cases, the transformation is "exact" in the sense previously defined for MOVE, and completely **local** in space, i.e. each lattice site can be updated independently of all the others. As we shall see, this property is key to the success of LGCA as a parallel algorithm for CFD.

By combined application of streaming and collision, one derives the following evolution equation for LGCA:

$$n_i(\vec{x} + \vec{c_i}, t + 1) - n_i(\vec{x}, t) = C_i(n) \qquad i = 1, b \tag{4.8}$$

where b is the number of bits per grid-point and $\vec{c_i} = \Delta\vec{x_i}/\Delta t$ are the discrete particle speeds in lattice units ($\Delta\vec{x_i} = \Delta t = 1$), $\Delta\vec{x_i}$ being the connectivity vectors linking each node to its neighbors.

How does the LGCA eq. 4.8 connect to the fluid equations ? This task is accomplished by defining mean macroscopic quantities like the fluid density ρ and speed $\vec{u}$, via an ensemble averaging over many histories, and then summing up over all discrete speeds. This leads to:

$$\rho(\vec{x}, t) = \sum_i < n_i > \tag{4.9}$$

$$\vec{u}(\vec{x}, t) = \sum_i \vec{c_i} < n_i > / \rho(\vec{x}, t) \tag{4.10}$$

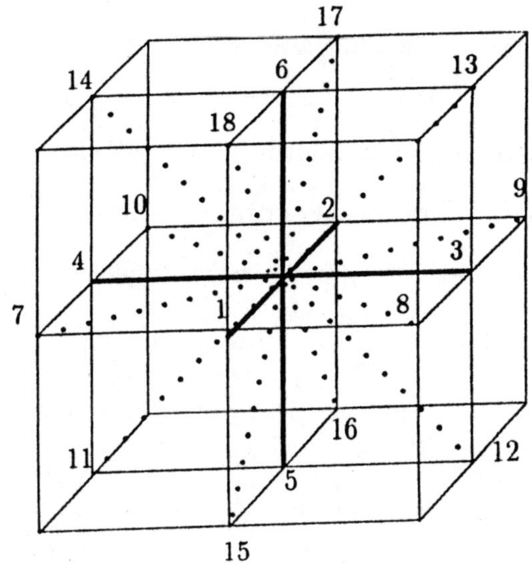

Figure 4.6: The 3D projection of the 4-dimensional FCHC lattice. Solid links carry two propagating particles with speed ± 1 respectively along the fourth dimension

In order for the macrovariables ρ and $\vec{u}$ to fulfill the Navier-Stokes equations, two basic conditions have to be met:

a) Mass and momentum must be locally conserved
$\sum_{i=1}^{b} C_i = 0$, $\sum_{i=1}^{b} C_i \vec{c_i} = \vec{0}$

b) The set of connecting speeds $\{\vec{c_i}\}$ must ensure rotational isotropy of the fourth order tensor $\vec{\vec{T}} = \sum_{i=1}^{b} \vec{c_i}\vec{c_i}\vec{c_i}\vec{c_i}$

where b is the number of discrete speeds that define the lattice connectivity. Condition b) is particularly demanding, in fact it restricts the class of lattices supporting hydrodynamic behaviour to a very few elements, the most popular being the hexagonal FHP lattice in 2 dimensions ($b = 6$) and the FCHC (face-centered hypercube) in four dimensions ($b = 24$) see fig. 4.6. Three dimensional hydrodynamics is obtained by projecting FCHC onto a three-dimensional space.

LGCA are sometimes referred to as **M**aximally **D**iscretized **M**olecular **D**ynamics, in the sense that they correspond to the most radical move in the direction of replacing true Newtonian mechanics with a stylized surrogate still preserving the key physical ingredients of a fluid (i.e. conditions a) and b)). This drastic simplification results in a series of pro's from the computational point of view:

+1 Exact computability (immunity to round-off errors)

+2 High spatial-temporal locality (strong potential for parallel computing)

+3 Ease of incorporation of grossly irregular geometries (porous media)

These advantages are however counterbalanced by a number of con's:

-1 Statistical noise: many histories are needed to extract smooth hydrodynamics out of LGCA dynamics.

-2 Exponential complexity of the collision map: as more physics (three-dimensional effects, thermodynamic phenomena,...) is included, due to the high number of required speeds.

-3 Topological stiffness: at variance with MD, in which particles are free to move around in space, LGCA are bound to "live" in a limited class of lattices with rigid symmetry specifications (conditions a) and b))

LGCA were hailed as a real breakthrough in the late 80's, mainly in view of +1) and +2); later on, this enthusiasm has somehow waned out, mostly on account of -2) and -1) respectively.

Still, LGCA have made their way into some niche CFD applications such as creeping flows in grossly irregular geometries with possibly accompanying heteregeneous chemical reactions [65], [66].

Besides, some researchers do believe that by resort to special-purpose customized "lattice gas processors", LGCA may outperform current leading-edge computers by a few orders of magnitude [67], [68]. Future only will make the final statement.

For the time being, at least on general-purpose machines, LGCA have been significantly superseded by a close relative, known as the Lattice Boltzmann method.

4.1.4 Lattice Boltzmann

Two major drawbacks of LGCA -1) and -2)) can be cured by a qualitative change in perspective: instead of propagating "hard bits" (either a particle is on the node or it is not: *tertium non datur*), one may decide to evolve the "mean probability" $N(\vec{x},t)$ for a particle to sit at node $\vec{x}$ at time t , with speed $\vec{c_i}$. This probability is nothing but the ensemble average of the boolean occupation number n_i: $N_i = < n_i >$.

From a methodological point of view, this means moving from the Newtonian level to the Boltzmann level of the BBGKY hierarchy.

Under the assumption of molecular chaos, i.e. no correlations between particles entering a collision, and small deviations from local thermodynamic equilibrium, ensemble averaging of the LGCA dynamics yields the following semi-linear evolution equation for the $N_i's$ (Lattice Boltzmann Equation -LBE-) [69] [70]:

$$N_i(\vec{x}+\vec{c_i},t+1) - N_i(\vec{x},t) = \sum_{j=1}^{b} A_{ij}(N_j - N_j^{eq}(\rho,\vec{u})) \tag{4.11}$$

Here, A_{ij} is a scattering matrix accounting for interactions between N_i and N_j, and N_j^{eq} is a local equilibrium distribution which depends parametrically on the local hydrodynamic conserved quantities $\rho(x,t)$ and $\vec{u}(x,t)$.

A typical form of N_j^{eq} suitable for hydrodynamics is

$$N_j^{eq} = \rho(1 + \alpha\vec{u}\cdot\vec{c_i} + \beta \overset{\Rightarrow}{Q}_i : \vec{u}\vec{u}) \tag{4.12}$$

where α, β are adjustable coefficients and $\overset{\Rightarrow}{Q}_i$ is the lattice projector along the i-th speed:

$$\overset{\Rightarrow}{Q}_i = (\vec{c_i}\vec{c_i} - c^2/4 \overset{\Rightarrow}{I}) \tag{4.13}$$

where $\overset{\Rightarrow}{I}$ denotes the unit tensor.

Hydrodynamic behaviour of $\rho(x,t)$ and $\vec{u}(x,t)$ under the evolution of eq. 4.11, the Lattice Boltzmann Equation (LBE), is ensured under the same symmetry and conservation requirements of LGCA dynamics.

For LBE, conservation requirements translate into the following sum rules for the scattering matrix:

$$\sum_i A_{ij} = 0; \qquad \text{(mass conservation)} \tag{4.14}$$

$$\sum_i A_{ij}\vec{c_i} = \vec{0_j}; \qquad \text{(momentum conservation)} \tag{4.15}$$

We anticipated that LBE was generated in response to the problems -1) and -2) affecting LGCA, i.e. statistical noise and exponential escalation of the complexity with the number of speeds/grid point. The reason why statistical noise is no longer a problem with LBE is quite clear: since the $N_i's$ are mean quantities thet are **by definition** noise-free.

Exponential complexity is lifted because the collision operator of LBE only entails b^2 (instead of 2^b) operations, which makes a big difference if b is sufficiently large !

The reason behind this dramatic reduction is the fact that b-body collisions, the cause of exponential blow-up, can safely be discarded for $b > 2$ as long as large-scale hydrodynamic behaviour is concerned.

As to the third inconvenient of LGCA, topological stiffness, LBE is exactly in the same position as LGCA: very special lattices are needed to recover hydrodynamic behaviour, unless one combines LBE with standard grid techniques.

Coming to the advantages +1),+2),+3) of LGCA, the only one which is lost along the way by LBE is "exact computability", since, as we have seen, boolean variables have been dismissed to allow a noise-free formulation. As a result, the (almost) ideal amenability to parallel computing characterizing LGCA is integrally inherited by LBE.

Although decidedly less revolutionary in nature than LG, LB has met with significant success within the CFD-oriented physics community, mostly in connection with simulation of fluid flows involving complex physics (multiphase, chemical reactions in heterogeneous environments) to cite but a few [71].

4.1.5 Vortex methods

We have already stressed the idea that the concept of "particle" should be intended in a very broad sense, as a discrete entity representing a "quantum" of a given physical quantity. Depending on the regime one is interested in, "particles" can be attached to different physical properties, such as mass, momentum energy and others. We also have already mentioned that the dynamics of high-Reynolds flows is most conveniently described in terms of the vorticity field

$$\vec{\omega} = \vec{\nabla} \times \vec{v} \tag{4.16}$$

which obeys the eq. of motion 2.41.

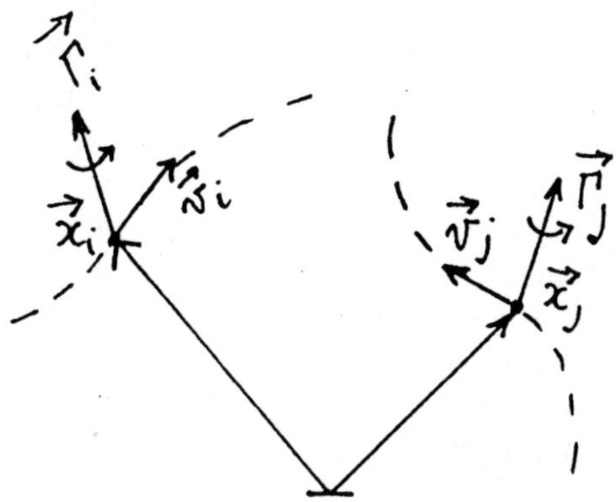

Figure 4.7: Two vortons sitting at $\vec{x}_i$, $\vec{x}_j$ with speeds $\vec{v}_i$, $\vec{v}_j$ and angular momenta $\vec{\Gamma}_i$, $\vec{\Gamma}_j$.

It is therefore quite natural to seek for discrete representations of fluid flows in terms of quanta of vorticity, or, as they are usually called, "vortons", i.e. point-like structures carrying a finite amount of angular momentum [72].

A system of N_V vortons located at $\vec{x_i}$ with angular momentum $\vec{\Gamma_i}$ gives rise to the following representation of the vorticity field:

$$\vec{\omega}(\vec{x},t) = \sum_{i=1}^{N_V} \vec{\Gamma_i}(t)\delta(\vec{x} - \vec{x_i}(t)) \tag{4.17}$$

The equations of motion for the 6 vorton coordinates $(\vec{x_i}(t), \vec{\Gamma_i}(t))$ are readily derived from the evolution equation of the vorticity field, complemented by the definition of the vorticity (eq. 2.35).

Straightforward algebra yields, for an inviscid fluid:

$$\begin{cases} \frac{d\vec{x_i}}{dt} = \vec{v_i} = \sum_{j=1}^{N_V} \frac{\vec{\Gamma_j}}{2\pi} \frac{\vec{r_{ij}}}{r_{ij}^2} \\ \frac{d\vec{\Gamma_i}}{dt} = \sum_{j=1}^{N_V} \frac{\vec{\Gamma_i} \cdot \vec{\Gamma_j}}{2\pi r_{ij}^3} \vec{r_{ij}} \qquad \vec{r_{ij}} = |\vec{x_i} - \vec{x_j}| \end{cases} \tag{4.18}$$

This is an hamiltonian system which bears close analogy with a system of electrically charged particles interacting via a **long-range** electrostatic, $1/r$, potential.

The system exhibits a quadratic non-linearity reflecting the vortex-stretching term, $(\vec{\omega} \cdot \vec{\nabla})\vec{v}$, which is responsible for vorton generation/annihilation processes.

Dissipative effects are accounted for by allowing random collisions as vortons get sufficiently close one to another. These random collisions redistribute fluid vorticity in space thus giving rise to effective diffusion which allows for vortex breakup and reconnection.

The vortex method is computationally appealing because, like MD, is virtually grid-free. At variance with MD, however, it is also noise-free because $\vec{\omega}$ is a collective, macroscopic variable.

Among the potential pitfalls, we quote the difficulty of implementing no-slip boundary conditions, for which cumbersome tricks based upon the introduction of "image charges" are needed, as well as some lack of robustness in terms of convergence to a smooth hydro-field as the number of vorton NV is increased [73].

Finally, the fact that vortons experience a long-range, **non-local** potential calls for specialized techniques to beat the N^2 complexity of the vorton "collision" term on the r.h.s. of the eq. 4.17 [74].

4.2 Smoothed Particle Hydrodynamics

Smoothed Particle Hydrodynamics (SPH) is a method for solving continuum equations with particles playing the role of moving grid points [75], [76].

Particle trajectories are dictated by the underlying continuum partial differential equations, while continuum fields, density, velocity and energy are represented as a superposition of individual particle contributions.

The peculiarity of SPH is that particles are no longer point-like objects, but their contribution to the continuum fields is mediated by a weight, or shape, function $W_h(r)$ extending over a finite spatial extent h.

Given a (non-smooth) fine-grained microscopic quantity $f(x,t)$, the corresponding SPH-filtered, coarse-grained (smooth) macroscopic field $F_h(x,t)$ is given by:

$$F_h(x,t) = \int W_h(|x - x'|)f(x',t)dx' \tag{4.19}$$

From this expression we quickly recognize an intimate link between SPH and (dynamic) interpolation theory.

Actually, since the "particle size" h can be made itself a function of space-time, the SPH method can naturally accomodate **multiscale** interpolation.

Once the basic constraints $\int W(x)dx = 1$ and $\lim_{h \to 0} W_h(x - x') = \delta(x - x')$ are fulfilled, the specific form of the weight function is a matter of choice related essentially to the desired degree of smoothness of $F_h(x,t)$.

A popular choice in fluid-dynamics application is the so-called Lucy function:

$$W_h^{Lucy}(r) = \begin{cases} C_L(1 + \frac{3r}{h})(1 - \frac{r}{h})^3; & r \leq h \\ 0 & r > h \end{cases} \tag{4.20}$$

where C_L is a normalization constant.

An intriguing aspect of SPH is that the weight function bears more than a superficial resemblance to the pair potential $V(r)$ familiar from Molecular Dynamics.

To see this in a more detail, let us consider the case where the flow variables, density, velocity and pressure, are known at a finite number of discrete points $\{X_i; \quad i = 1, N\}$.

Simple calculations show that the SPH representation of the continuity equation plus the Euler equation for inviscid flows reads as follows:

$$\begin{cases} \frac{dX_i}{dt} = V_i \\ \frac{dV_i}{dt} = \sum_j (\frac{P_i}{\rho_i^2} + \frac{P_j}{\rho_j^2} m_j \frac{\partial}{\partial x_i} W_h(x_i - x_j) \end{cases} \tag{4.21}$$

where ρ_i, V_i, P_i are the SPH representation of the density, velocity and pressure fields at $x = X_i$.

It is readily recognized that for the case of a two-dimensional expanding gas ($p \propto \rho^2$), the eq. 4.21 is exactly identical to Molecular Dynamics for point-like particles subject to a pair potential $W(r)$.

Building upon this intriguing analogy, a number of researchers are currently investigating fundamental issues of statistical mechanics, like the link of SPH with the irreversibility of macroscopic equations [77].

SPH is still a fast-evolving subject, so that any conclusive assessment on its impact on fluid dynamics should be deferred by a few years. Still, at least from a formal point of view, it appears an ideal candidate for future "multiphysics" applications combining atomistic and continuum methods on massively parallel computers [78].

4.3 Time marching with particle methods

We shall briefly sketch here the main time-marching methods used with the various particles methods described in this book.

4.3.1 Molecular Dynamics

Molecular dynamics time-stepping invariably proceeds through explicit integrators, implicit time-stepping being ruled out by the fact that as particles move around, their neighborhood is changing too.

This means that the force matrix , not only changes its values, but can also undergo structural changes in its sparsity pattern.

One of the most popular MD time-marching scheme is the two-level Verlet algorithm [79]:

$$x_p(t+\Delta t) = 2x_p(t) - x_p(t-\Delta t) + \frac{F_p(t)}{m}\Delta t^2 \tag{4.22}$$

This is 3rd order accurate in Δt, and does not require the computation of particle velocitites v_p.

Whenever needed, these can always be recovered by a simple 2nd order difference:

$$v_p(t) = \frac{x_p(t+\Delta t) - x_p(t-\Delta t)}{2\Delta t} \tag{4.23}$$

Other popular alternatives are the Gear predictor-corrector [80] and the leapfrog [81] schemes.

4.3.2 Lattice Gas and Lattice Boltzmann

As we have seen in Chapter III, LG and LB can be viewed as a sort of maximally discretized molecular dynamics, in the sense that the velocity degrees of freedom are drastically reduced to a few possible discrete speeds $\{\vec{c_i}\}$.

This drastic discretization implies that at each time-step, particle are bound to lie on the nodes of the discrete lattice generated by the $\{\vec{c_i}\}$'s $\vec{c_i} = \vec{a_i}/\Delta$, where $\vec{a_i}$ are the oriented links connecting neighboring lattice sites along the i-th direction.

The result is that discrete space and time are no longer independent and particle proceed in a **fully synchronous** lock-step mode corresponding to an explicit scheme working at the edge of the CFL condition:

$$|\vec{c_i}\Delta t|/|\Delta\vec{x_i}| = 1$$

In the case of the LB the scheme is stable provided the leading non-zero eigenvalue of scattering matrix, λ, falls within the interval $[-2, 0]$. Physically, this corresponds to a CFL condition on the mean-free path:

$$\frac{l_{mfp}}{\Delta x} \ll 1$$

and ensures that we reach hydrodynamic equilibrium within a single cell.

4.3.3 Direct Simulation Montecarlo

DSMC also proceeds via explicit time marching, with the time-step size fixed by a CFL condition

$$\left|\frac{v\Delta t}{\delta}\right| < 1; \quad \delta = \min(\Delta x, l_{mfp}$$

Here v is a typical particle speed, l_{mfp} the mean-free path and Δx the linear size of the spatial cell. This CFL is specific to kinetic theory because it means that the timestep should not exceed the typical time lapse l_{mfp}/v between two particle collisions.

4.3.4 Vorton methods

As anticipated, the dynamics of high-Reynolds flows can be conveniently described in terms of "vortons", i.e. quanta of vorticity.

These vortons behave like spinning particles endowed with 3 spatial coordinates $\{\vec{x_v}, v = 1, N_V\}$ attached to their position and 3 coordinates $\{\vec{\sigma_v}, v = 1, N_V\}$ associated to their rotational momentum.

The vortons see each other via a long-range $1/r$ potential and give rise to a typical N-body problem with many similarities with gravitational problems.

The temporal tracking of these systems is usually performed by high-order Runge-Kutta integrators.

In the case of incompressible flows, a major concern is to prevent time discretization errors from spoiling the divergence-free condition $\text{div}\vec{v} = 0$ in the course of long-time integrations.

To this purpose, relaxation procedures are adopted, too specialized to be described here, which introduce the need of solving linear algebra problems for vorton angular momenta.

These are currently handled via highly underrelaxed (the matrix A is normally non-diagonal dominant) Gauss-Seidel schemes [82].

4.3.5 Smoothed-Particle Hydrodynamics

The key ingredient of SPH is the weight function $W(r)$ mediating between fine-grained fluctuating quantities and coarse-grained smooth hydrodynamic fields.

As discussed in section 4.1.5, the weight function is to a certain degree, matter of choice. This makes SPH somehow exposed to numerical instabilities. For this reason, SPH equations are usually advanced with 4th order accurate Runge-Kupta time-marching schemes.

Chapter 5

Parallel computing

Traditional computer architectures still work on the principles laid down by John von Neumann back in the early 50's. The von Neumann computer consists of four basic functional units [83]:

- A central processing unit (CPU)
- Memory unit (M)
- Input/Output devices (I/O)
- A Control Unit (CU)

These units reflect the three basic steps that have to be accomplished in order to execute a program, i.e. manipulate and process **data** according to a prescribed set of rules (**instructions**):

1. Load data and instructions from Memory to CPU
2. Process data within the CPU
3. Store data back from CPU to Memory

Every now and then data need to be transmitted to the exterior in some form (printed output, graphic displays, disks, etc.), a task which is accomplished by peripherical devices (I/O unit).

For scientific applications the most common operations are (almost) invariably floating-point (flop) operations, i.e. add, subtract, multiply and divide (+, -, *, /).

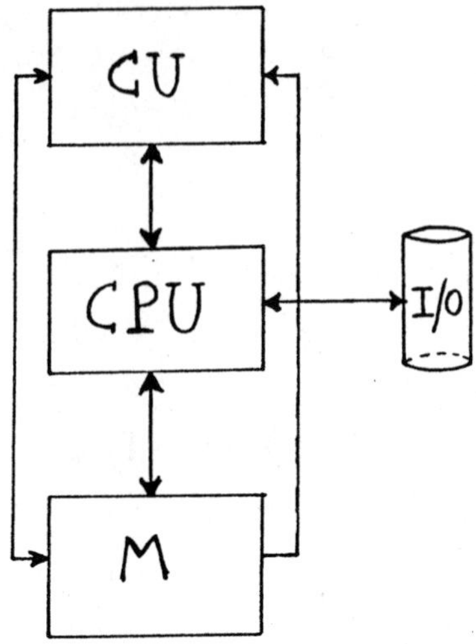

Figure 5.1: The basic units of the von-Neumann computer

This is why computer performance, in terms of raw processing speed, is most frequently measured in number of floating-point operations per second (**flop/s**).

Assuming for a while that F flop/s can be executed in a single clock-time, the processing speed is simply given by the inverse clock-time times F:

$$\mathcal{S}[flop/s] = \frac{F}{\tau_c} \tag{5.1}$$

where τ_c is the clock-time in seconds. As of today, computer technology can afford clock-times as low as a few nanoseconds, which means a theoretical peak speed of the order of $Gflop/s$ (a billion flop/s). Fantastic as it is, this figure is still far from being enough to satisfy the appetite of many computational applications, CFD included.

To convince oneself, think of a full CFD simulation at $Re \simeq 10^6$ (moreless the case for a standard cruisade jet). According to eq. 1.2, this requires about $(10^6)^{9/4} \simeq 3 \cdot 10^{13}$ degrees of freedom. Assuming each degree of freedom takes 300 flop to be updated in a single time step, this means 10^{16} = 10 million Gflop/time-step: about half a year to compute just a single time-step !

In spite of the progress of computer technology, it appears like engineering (rather than physical) constraints will set the ultimate barrier for shorter and shorter clock times not much farther than today's standards [84].

Perhaps, the most stringent constraints are those related to thermal evacuation issues: two chips working at 0.1ns clock-time cannot be placed farther than 3cm apart (distance travelled by a light signal in 0.1ns). As a result, clock-time reduction is behindered by the concurrent effect of two adverse factors: higher heat production per chip (fast switching requires more power) and higher chip density (to offset propagation delay times).

That's why, according to Seymour Cray, "computer architects are basically cooling engineers".

It appears therefore that a much more viable strategy to increase computer speed consists of letting **several** processors to cooperate simultaneously to the solution of a given problem by working on distinct subportions in an independent fashion.

Calling P the number of processors in point, this strategy offers a potential gain of a factor P in effective processing speed.

In practice, however, many factors arise which prevent such a linear speed-up from being attained. As a result, we shall more realistically assume that the processing speed with P processors is given by:

$$\mathcal{S}(P) = \eta(P)P\frac{F}{\tau_c} = \eta P\mathcal{S}(1) \tag{5.2}$$

where $\eta(P)$ is the **parallelization efficiency**, a number usually (but not necessarily !) between zero and one, that depends on the type of parallel architecture as well as on the details of the computational algorithm.

The aggregation of multiple processors corresponds to what we may label as **"inter-processor"** parallelization, in the sense that the single processor is left untouched and higher speed obtained via sheer multi-processor aggregation.

A second possibility is **"intra-processor"** parallelism, which consists of equipping the single CPU with special units aimed at raising the number of flop per cycle, F.

This is the basic idea behind **vector** computers, and, more recently, **superscalar** architectures, which shall be described in a little more detail in the sequel.

For the time being, we simply reiterate the basic belief behind parallel computing, i.e. **is technologically more viable to increase S by bundling together many processors rather than making the single processor increasingly faster**.

It should be appreciated that this philosophy implies a strong shift from hardware issues related to the physics and technology of semiconductors, to software issues connected with the optimal management of complex computer networks.

Any means of achieving higher performance without lowering the clock-time, must draw upon some form of concurrency.

Let us first examine the case of "internal parallelism" and discuss the two most popular architectures relying upon it: vector computers and superscalar computers.

5.1 Vector computers

Vector computers are machines designed to achieve top processing speeds by pushing clock-times as low as possible and targeting at the same time the goal of one flop/cycle [85] [86].

The former goal is pursued via sheer technological means, while the latter rests upon an architectural strategy known as "pipelining".

On a traditional computer, the completion of a floating-point operation involves the execution of a certain number of sequential sub-operations. For instance, in the case of the addition of two real numbers x, y to produce the sum $s = x + y$, the following steps need to be performed:

1. Compare exponents
2. Mantissa alignment
3. Mantissa addition
4. Normalization of the sum

This takes 4 cycles, so that $F = 1/4$.

It is useful to view this process as the operands x, y going through a 4-stage **pipeline** to undergo the transformation needed to produce the results.

Suppose now that the **same** operation, the addition in our case, is to be performed on a whole array (vector) of N numbers $x = (x_1, x_2, ..., x_N)$, $y = (y_1, y_2, ..., y_N)$ to produce an array of sums $s = x + y$.

Under these conditions, it is clear that the goal of one result per cycle ($F = 1$) can be achieved by organizing the addition as a sort of assembly-line, whereby the various sub-operations 1 to 4 are executed concurrently on different data items.

This technique, known as **pipelining**, is able to deliver one addition per cycle after a certain start-up time of the order of k cycles, where k is the pipe length, provided the vectors are long enough, i.e. $N \gg k$ (Fig. 5.3).

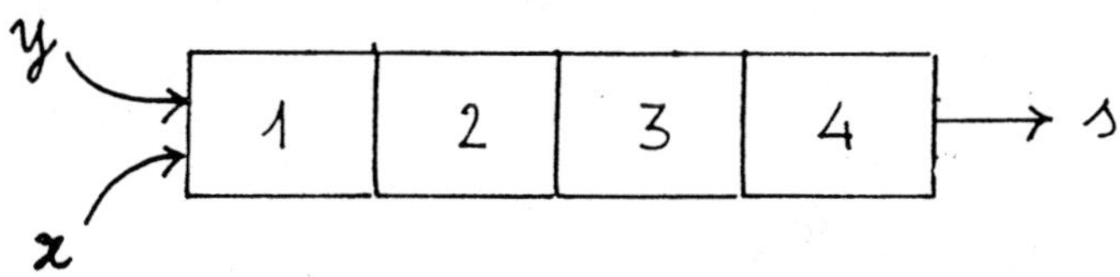

Figure 5.2: The operational scheme of pipelining. Note the close analogy with an assembly line

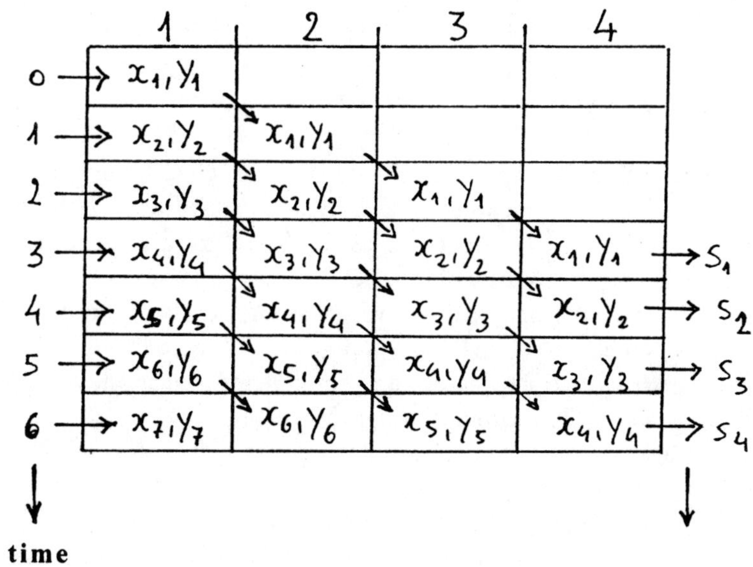

Figure 5.3: The operational principle of pipelining

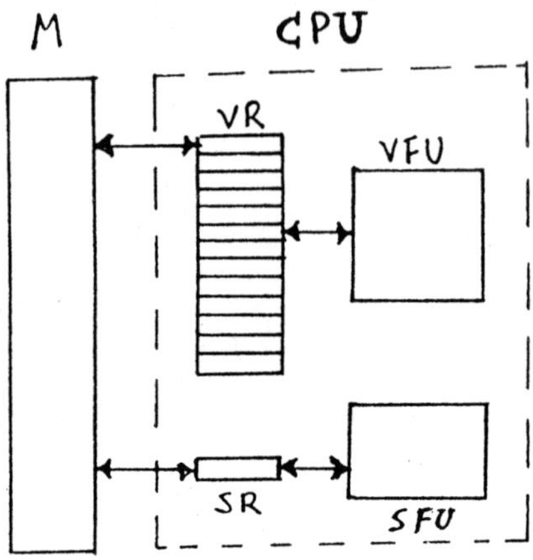

VR; vector registers

SR; scalor registers

VFU; vector functional units

SFU; scalar functional units

Figure 5.4: The basic functional units of a vector architecture

More precisely, by neglecting the cost of memory access, a vector computer completes the N additions in $(N + k)$ cycles, as opposed to the Nk cycles needed by a scalar one. This yields a vector speed-up of

$$S_v = \frac{Nk}{N+k} = \frac{k}{1+k/N} \sim k \quad \text{for } N \gg k \tag{5.3}$$

It is clear that in order for this speed up not be spoiled by memory access delays, vector CPU's must cater for sufficient local storage in the form of **vector registers** holding the operands as well as the results. The schematic of a vector CPU is indicated in Fig. 5.4:

The ability of a vector architecture to amortize the overhead due to memory access is measured by the "$N_{1/2}$ parameter" introduced by Hockney and Jesshope [87], which expresses the vector length such that half of the full speed is achieved. The smaller $N_{1/2}$, the better.

Typical behaviour of the computing speed as a function of the vector length N is represented in Fig. 5.5

For most modern vector architectures $N_{1/2}$ is in the order of $10 - 20$

Finally, we mention the remarkable possibility of some vector machines to perform compound operations in a single cycle (for instance the "saxpy" operation: $y_i = y_i + a * x_i$). This is obtained by the mechanism of chaining, i.e. the output of a pipe serving as an input to next pipe:

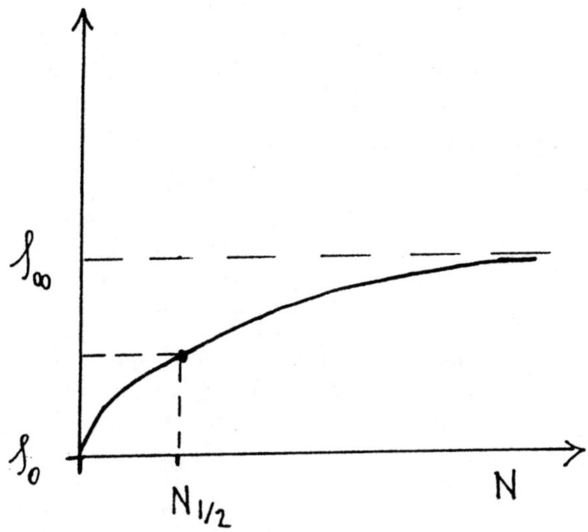

Figure 5.5: Qualitative behaviour of the computing speed as a function of the vector length N

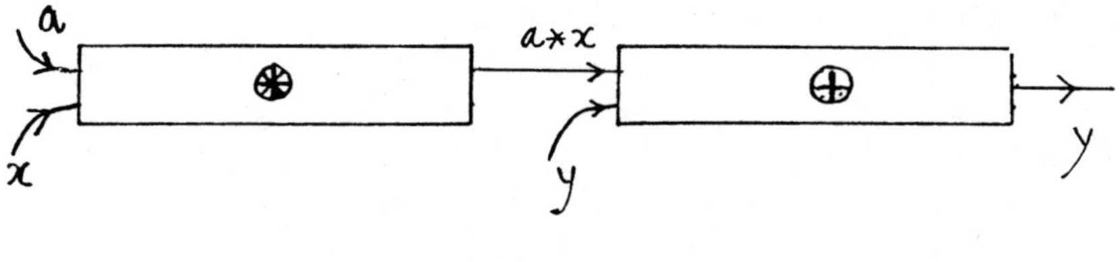

Figure 5.6: The basic principle of chaining. The output of the first pipe serves as input of the second to produce the *saxpy* result $y = y + a * x$

5.2 Superscalar computers

Since the inception of the first Cray-1 (1972), vector supercomputers have been holding the lead of large-scale scientific and technical computing.

In the recent years, basically on account of the dramatic advances of CMOS (**C**omplementary **M**etal **O**xide **S**emiconductor) technology, the leadership of vector computers has been seriously threatened by RISC (**R**educed **I**nstruction **S**et **C**omputer) processors such as those commonly employed in workstations. Indeed, on many scientific applications, RISC-based computers feature superior performance per dollar as compared to traditional supercomputers [88].

Altough CMOS technology is intrinsecally slower than the one used for vector machines, (**E**mitter **C**oupled **L**ogic), its superiority in terms of chip packaging densities has given access to fairly comparable clock rates (in excess of 100 MHz, as compared to 166 MHz of the Cray Y-MP/1, or 240 MHz of Cray C-90/1).

In addition, by capitalizing on RISC architectural simplicity, sophisticated compilers are available which can raise the flop/cycle as high as for vector supercomputers [89].

Such a high flop/cycle ratio is the result of a special type of concurrency which takes place at the **instruction level**: due to the presence on the same chip of independent functional units, multiple instructions (add, multiply,...) can be issued and executed simultaneously, whence the label of **superscalar architectures**.

A typical sketch of a superscalar RISC processor is presented below:

A superscalar CPU that can execute multiple instructions per cycle requires high instruction and data bandwidths to be kept busy.

In the IBM RISC System/6000, for instance, there are a 4-word memory bus, a one-word data bus between the FiXed point Unit (FXU) and data cache (D-cache), and a 2-word data bus between Floating Point Unit (FPU) and D-cache.

Caching is a distinctive feature of RISC design: a cache is a "fast memory" buffer designed to retain a moderately small amount of data and instructions (hundreds of kBytes) which can be read-off at the rate of one word/cycle.

Caches work on the principle of **data locality**: the assumption is that if a data item, say $A(i)$, is needed at a given time, the items that are more likely to be needed at subsequent times should be contiguous to $A(i)$. Hence, when $A(i)$ is requested, a whole bunch of contiguous data ("cache line") is

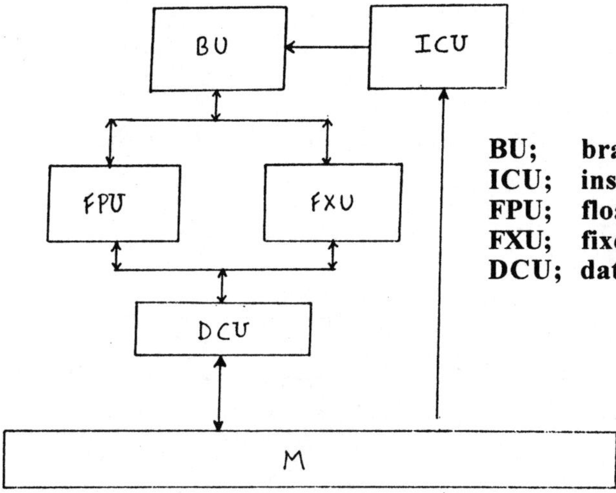

BU; branch unit
ICU; instruction cache unit
FPU; floating point unit
FXU; fixed point unit
DCU; data cache unit

Figure 5.7: A typical layout of a superscalar architecture. Note the presence of multiple functional units

loaded into the cache for future use.

Algorithms featuring high "cache-hit ratios" (memory references successfully served by the cache) can come very close to peak performance.

High cache-hit ratios are (relatively) easily obtained in one and two dimensions, but much less in three, due to the intrinsic occurrence of large **strides** (memory offset between two subsequent references).

Take for instance two memory references such as $A(i,j,k)$, $A(i,j,k+1)$; $i = 1, NX; j = 1, NY; k = 1, NZ$. Although physically contiguous, these two elements are separated by $NX \cdot NY$ words in computer storage.

Several techniques to enhance data locality and leverage the features of superscalar machines, too detailed to be described here, are available in the specialized literature [90].

5.3 Parallel architectures

As we already pointed out, the approach of enhancing the sheer technology inside the computer seems to be reaching its ultimate limits, in the sense that the laws of physics and especially engineering will inhibit future progresses in a reasonable time at affordable costs.

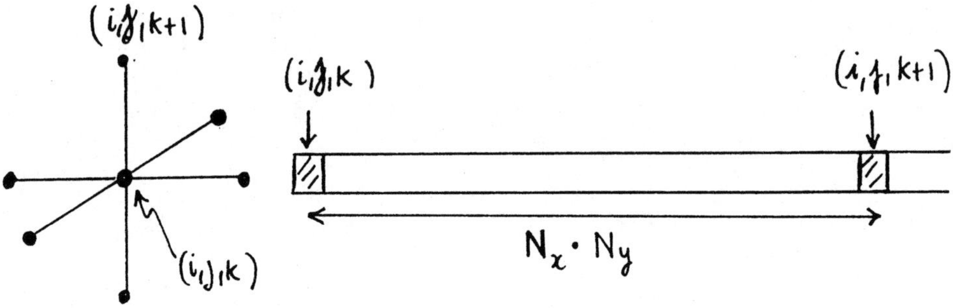

Figure 5.8: Physically contiguous data map into logically distant memory locations

To overcome these restrictions, the last decade has witnessed the emergency of parallel machines, i.e. computers in which the uniprocessor is relieved from doing all the work by introduction of multiple processors in a single machine.

Parallel machines come into many families, depending on the following main features:

- Control mechanism
- Memory organization
- Interconnecting topology
- Granularity

5.3.1 Control mechanism

As early as 1972, Michael J. Flynn conceived a taxonomy of parallel machines which is still widely popular today. In this classification an **instruction stream** is defined as the sequence of instructions executed by a single processor, while a **data stream** pertains to the data instructions act upon [91].

Flynn identifies the following four categories:

SISD (Single Instruction Single Data)

This is basically the traditional Von Neumann architecture.

SIMD (Single Instruction Multiple Data)

This category includes so called **array processors**. Here a single control unit (CU) directs the sequence of operations. It broadcasts one instruction at a time to all processors which perform the same instruction in a synchronous lock-step mode.

MISD (Multiple Instruction Single Data)

Here multiple instructions are executed on the same data. This looks quite innatural, for physics seems to proceed the other way around: the same (a few) laws acting on many different data. And indeed, no MISD implementation exists for interprocessor parallelism. Yet, the idea does just fine for intra-processor parallelism; this is precisely the principle that vector computers are based on.

MIMD (Multiple Instruction Multiple Data)

This class includes **multiprocessor** computers and comprises most of today's parallel architectures. The typical MIMD architecture is represented in Fig. 5.10.

Here each CPU can execute a different instruction, thus resulting in a much more flexible architecture allowing great generality of the programming model.

5.3.2 Memory organization

This classification differentiates between systems in which all processors share a common memory (**tightly coupled**) and those where each processor owns its local memory (**loosely coupled**), or **distributed memory**.

Tightly coupled (or **shared memory**) systems usually employ a relatively small (up to a few tens) number of powerful, expensive processors. Most of today's vector multiprocessors belong to this class.

Since all processors see a single memory space via a common bus, interprocessor communication is "transparent" to the programmer. High efficiency can be obtained provided memory contention between the various processors (two or more processors issuing requests to the same memory location) is properly arbitrated.

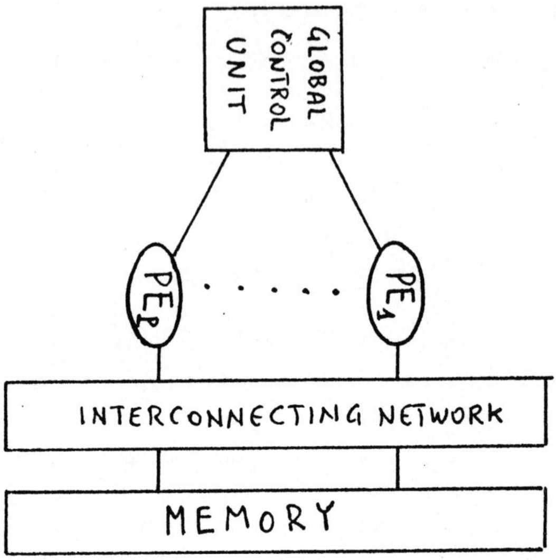

Figure 5.9: Sketch of a SIMD architecture

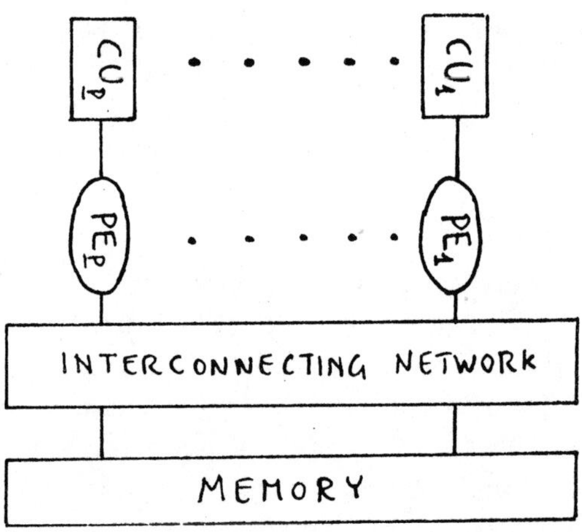

Figure 5.10: Sketch of a MIMD architecture

This is controlled via software through a **locking** mechanism, whereby a given processor gets the (exclusive) right to access and modify data that will be subsequently accessible to other processors.

The efficiency of this locking mechanism decreases quite fast with increasing number of processors due to the (exponential) growth of potential memory contentions for highly-communicative applications.

At the same time, the hardware complexity required to guarantee uniform memory access time over a pool of P processors grows like P^2, thus rapidly leading to unmanageable situations if P becomes too large ($P \geq O(10)$).

These two drawbacks are commonly agreed to hamper the **scalability** of shared-memory multiprocessors, i.e. their ability to draw gains of performance in proportion to the number of processors, for more than a few tens of processors.

In **distributed memory** machines, each processor owns and has exclusive access to its private local memory. Interprocessor communication takes place via explicit message exchanges across an interconnecting network.

Other interprocessor communication paradigms can be conceived, such as implicit message-passing or global shared-address memory, but we shall not be concerned with these in the present book.

The performance of distributed memory architectures is strongly dependent on the main technological features of the network, i.e.

- topology (how processors are connected)
- latency (minimum time required to establish the connection between two processors)
- bandwidth (maximum rate at which data can flow from one processor to another once the link is established)

At the application level, a key role is played by nature and the communication software [92]. While for shared memory machines efficient parallelizing compilers already exist, the same is not true for distributed memory machines.

5.4 Topology

Topology defines the way that the various processors are interconnected; in other words who are the neighbours of a given processor. This notion has a definite impact on the performance of the architecture because it's clear that neighbor-communication is privileged in that it is served via hardware, as opposed to remote, non-local, communication for which software mechanisms are required.

In fact, whenever long-distance (non-neighbor) communication is required between, say processor P_i and P_j, a software mechanism (**routing**) must be invoked to select the most convenient path linking P_i to P_j via intermediate neighboring connections

$$P_i \rightarrow P_j = P_i \rightarrow P_{i1}\ P_{i1} \rightarrow P_{i2} \ldots\ldots P_{iN} \rightarrow P_j$$

Ideal performance results whenever the communication pattern is homeomorphic to the computational stencil, i.e. the topology of the network reflects exactly the one dictated by the numerical scheme.

This trivializes an otherwise fairly complex problem known as *problem-to-processor mapping*, that is to say the optimal assignement of computational tasks to the various processors.

A typical example is the 4-node Laplacian discretization in two-dimensions on a regular Cartesian grid of unit mesh size:

$$\Delta\Phi \rightarrow \Phi_{i+1,,j} + \Phi_{i-1,,j} + i\Phi_{i,,j+1} + \Phi_{i,,j-1} - 4\Phi_{i,,j}$$

This expression is perfectly suited to a mesh-networked computer in which each processor, P, has direct connections to its "north-south" and "east-west" neighbors.

Computer networks split into two broad categories [93]:

- Static
- Dynamic

Static networks are those in which the connectivity pattern is fixed once and for all. Dynamic networks, on the contrary, support time-changing connectivity patterns.

Among the most popular static topologies we cite:

- Ring and one-dimensional torus
- Mesh and two-dimensional torus
- Hypercube

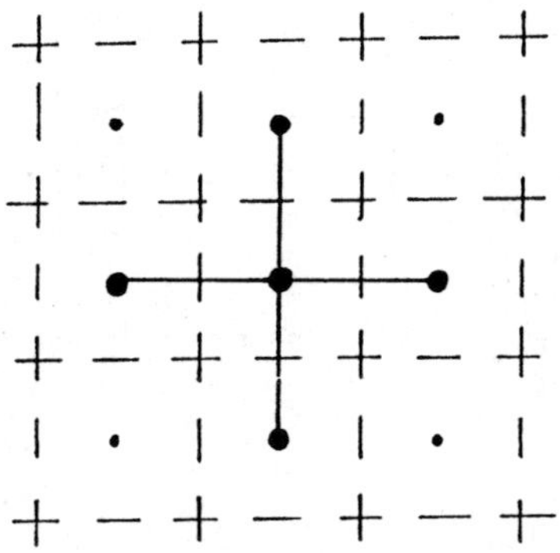

Figure 5.11: The computational stencil of the Laplacian in two-dimensions is naturally suited to a mesh network

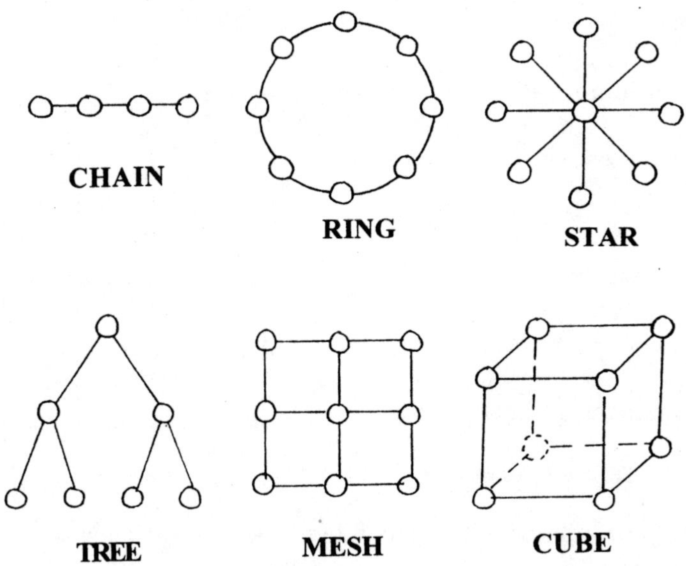

Figure 5.12: The main static-network topologies

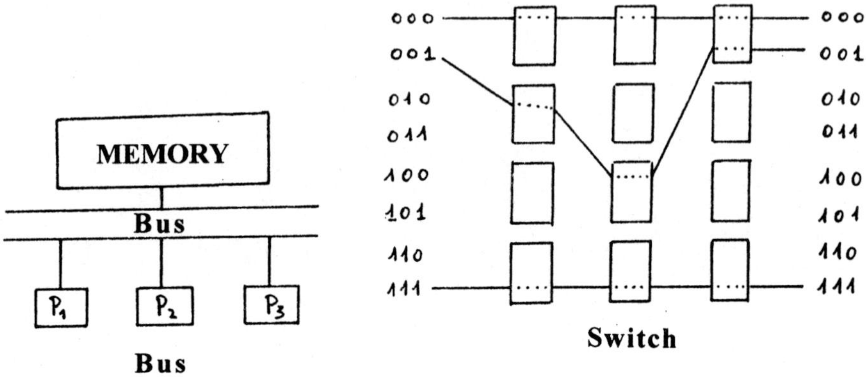

Figure 5.13: The main dynamic-network topologies

- Tree
- Fully connected

The most important dynamic networks are

- Crossbar switch
- Bus
- Multistage switch

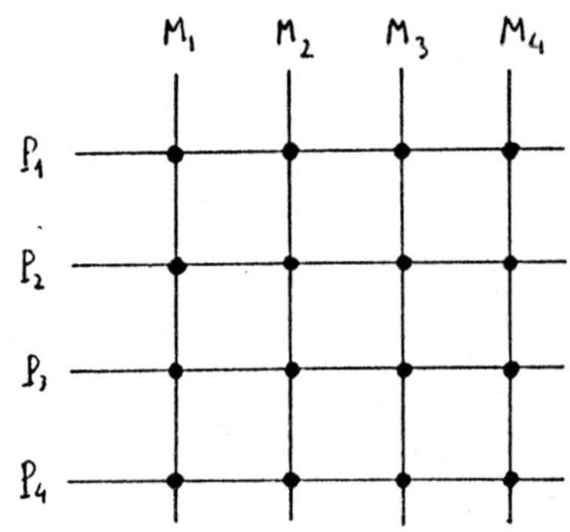

Crossbar

Ideally, a static network should feature:

- Low number of links per node
- Small internode distances
- High fault tolerance (high number of alternative paths between any two nodes)
- Good modularity (capability to increase the number of nodes without changing the topology)

For instance, the single ring topology has only two links/node but a high maximal internode distance, i.e. $N/2$, N being the number of nodes. In addition, in case of fault, the message is routed in one direction only, thus increasing the maximum internode distance, d_{max} to $N-1$. On the other extreme, a fully-connected network features $d_{max} = 1$, but at the price of as many as $(N-1)$ links/node.

The hypercube is a particularly interesting topology; it features $D = logN$ links/node ($N = 2^D$, where D is the dimensionality of the hypercube) and has $d_{max} = D$.

In addition, by a suitable numbering of the nodes, it can be taylored in such a way as to reproduce the structure of a mesh, tree or multiple hypercubes of lower dimensionality.

Let's come to dynamic networks. Crossbars support full connectivity because they provide direct links from any processor to any other. They don't scale however, because the number of links grows like the square of the number of processors.

Bus architectures are at the extreme opposite. They don't scale in connectivity since all processors share the same bus. They scale however in complexity because the bus complexity grows only linearly with the number of processors. Multistage switches represent a convenient compromise between the two.

Switched networks permit to establish connections between any two processors in the net. They are flexible in that the distance between any two processors is the same for any pair of processors so that the notion of topology is virtually erased. The limitation is that since each processor cannot have more than one mate at a time, one-to-many and many-to-one communications are partially serialized.

Which topology does best for fluid dynamics pretty much depends on the type of CFD methodology adopted.

For instance, it is clear that regular-uniform CFD schemes in D dimensions are ideally suited to D-dimensional meshes.

Hierarchical methods, such as multigrid, are instead ideally matched to tree-like networks.

The hypercube is particularly suited to 2D Fast-Fourier transforms because at each stage the network can be configured in such a way that neighboring data reside on neighboring processors.

Switched networks, owing to their topological freedom, can deal quite efficiently with unstructured irregular applications in which the neighborhood of a given node has no pre-defined structure and must be specified via indirect linked lists.

5.5 Parallel programming: general concepts

The main goal of parallel computing is three-fold [94]:

1. Reduce the elapsed time of a given application at a fixed size
2. Solve increasingly larger problems at a fixed cost in time
3. 1+2

Naively, the idea is to aggregate P processors to get a P-fold increase of processing speed:

$$\mathcal{S}(P) = P\mathcal{S}_1 \tag{5.4}$$

where $\mathcal{S}_1$ is the uni-processor speed.

In reality, many sources of inefficiency arise, some due to basic conceptual limitations, some others to practical architectural constraints, which prevent the ideal linear speed-up eq 5.4 to be achieved. As a result, the actual speed achieved by a P processor machine is more conveniently rewritten as

$$\mathcal{S}(P) = P\mathcal{S}_1\eta(P) \tag{5.5}$$

where η is the *parallel efficiency.*

The major challenge in parallel computing is the development of suitable numerical methods designed in such a way as to take full advantage of the potential gains offered by a parallel machine; in brief, achieve the highest possible efficiency.

Although far-deep details of a given algorithm might ultimately be tightly related to the machine architecture, there are a number of general concepts and guidelines to "efficient parallel programming" which hold their validity regardless of the specific architectural and application details [95].

The purpose of this section is to review these guideline concepts. Let us begin by listing the main issues controlling the parallel efficiency:

1. Fraction of parallel content
2. Load balance

3. Communicativity

4. Redundancy

Before discussing each of these points in detail, let us spend a few comments on the important issue of *scalability*.

5.5.1 Scalability

Scalability refers to the property of keeping the parallel efficiency virtually constant, $\eta \sim \eta_0$, across a whole range of processors (scaling region) $1 < P < P_c$. The wider the range and the closer η_0 to unit value, the better scalability is.

It is useful to express the efficiency in the following general form

$$\eta = \frac{1}{1+\omega}$$

where ω represents the "generalized" overheads incurred in by the parallel application to let the P processors exchange information and coordinate their joint operation so as to produce coherent results.

More precisely, ω expresses the **Communication to Computation Ratio** (CCR), i.e. the time spent into parallel overheads versus the time usefully spent to advance the calculation.

Focusing the attention on a single processor out of a pool of P, the computational cost generally decrease with the inverse of P, while the parallel overheads either increase or, at best, decrease more slowly than $1/P$. The ultimate result is that beyond a certain processor threshold, P_c, communication overheads take over and the efficiency undergoes a severe drop. This is where scalability begins to end.

Each specific type of overhead introduces its own P_c. As a result, it is often possible to recast the CCR in the following self-similar form

$$\omega = (\frac{P}{P_{c1}})^{a_1} + (\frac{P}{P_{c2}})^{a_2} + ...(\frac{P}{P_{cn}})^{a_n}$$

where $a_1, ...a_n$ are the *scaling exponents* associated to the various P_c's.

Clearly, the a's and P_c's depend on both the architectural and application parameters. Typically, the critical processors take also a self-similar form, such as:

$$P_{ci} = (\frac{N}{N_{ci}})^{b_i} \; i = 1, ...N$$

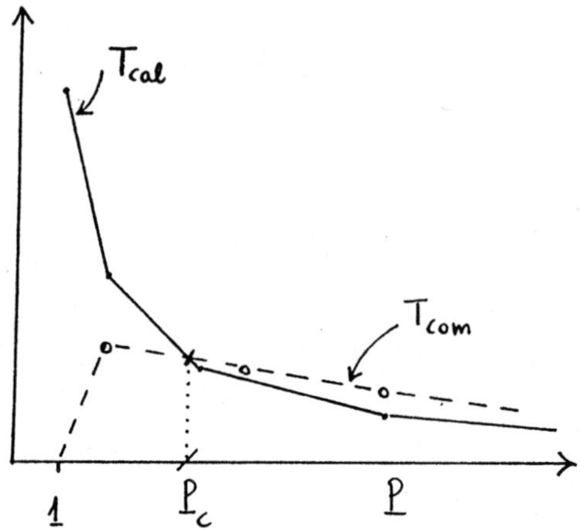

Figure 5.14: Qualitative behaviour of the communication and computation costs as a function of the number of processors

where N_{ci} is a typical problem size below which computation costs are shadowed by communication overheads. Note that the P_c's provide a quantitative indication on how the number of processors should be increased with the problem size in order to keep the efficiency to a constant value.

As a general remark, a good scalable application should feature high P_c's (wide scaling region) and small a's (gentle loss of scalability outside the scaling region).

Scalability can be examined under two distinct perspectives, the **continuum limit** and the **thermodynamic limit.**

They correspond to the following limits:

- Continuum: $\lim P \to \infty$, $N = const.$, $n = N/P \to 0$
- Thermodynamic: $\lim P \to \infty$, $n = const.$, $N \to \infty$

These two asymptotic regimes are by no means equivalent; indeed, the thermodynamic limit, corresponding to the job-upsizing scenario, assures much better scalability than the continuum limit, corresponding to the "cut turnaround time" scenario. The reason is obvious: the thermodynamic limit involves larger problems, hence higher potential for parallelization.

5.5.2 Fraction of parallel content

Most mathematical models of physical phenomena, including fluids, are computationally parallel up to a fairly large extent. Most enthusiast supporters of parallel computing tend to point out that there's a very solid reason for this: nature itself is parallel !

Atoms in a fluid don't line up to interact one each other (unlike in a computer program running on a serial machine), but just do it all together at the "same time".

Yet, nearly every program, be it running on a parallel machine or not, is made up by a mixture of **serial** portions, which must/can be executed by one processor at a time, and **parallel** portions which can be executed by each processor concurrently with all others.

A simple example illustrates the idea:

Suppose we're given the task of computing the sum S of N numbers $x_1, x_2, ..., x_N$ on a parallel machine consisting of $(P+1)$ processors. (for the sake of semplicity we shall assume that N is a multiple integer of P ; $N = k \cdot P$, k integer).

A fairly natural strategy consists of letting each of the P processors compute the partial sum of just k numbers, S_k, and then have all P-partial sums added together to produce the final result S.

Leaving aside input/output issues, this scheme involves two distinct tasks:

1. Compute partial sums S_p ; $p = 1, ..., P$
2. Sum partial sums $S = \sum_{p=1}^{P} S_p$

It is clear that while 1) can be split among P processors working independently on their portion of the problem, stage 2) presents work only for one processor ($p = 0$; or "master processor").

We assumed here, for the sake of simplicity, that after having completed its partial sum S_p, each processor in the pool $p = 1, P$ ("slave" processor) forwards its contribution S_p to the master processor, which is in charge of adding the partial sums to produce the final result.

As a result, we see that even though the topic of adding N numbers is intrinsically parallel, the actual program implementing this task consists nevertheless of a mixture of a serial (task 2) and a parallel (task 1) portion.

The presence of a serial portion of the code sets the upper bound for the speed-up which can be achieved by running a problem in parallel. This upper bound is easily computed as follows.

Denoting by f the fraction of CPU time spent in the parallel portion of the code, the ideal (minimum) time spent by using P processors is given by

$$T(f,P) = (1-f)T + f\frac{T}{P} \tag{5.6}$$

where $T \equiv T(f,1)$ is the time taken on a single processor. Hence the ideal speed-up, $S(f,P)$, is given by:

$$S(f,P) = \frac{T(f,1)}{T(f,P)} = \frac{1}{(1-f)+\frac{f}{P}} = \frac{P}{f+(1-f)P} \tag{5.7}$$

and the ideal efficiency

$$\eta(f,P) = \frac{1}{P}S(f,P) = \frac{1}{f+(1-f)P} \tag{5.8}$$

The eq. 5.7, known as Amdahl's Law in the context of vector computing, is just little more than common sense; yet it prompts to a series of important considerations [96].

The first remark is that even in the limit of an infinite number of processors, $P \to \infty$, the maximum achievable speed-up S_∞, is not infinite, unless the code is perfectly parallel, $(f = 1)$, indeed $S_\infty = (1-f)^{-1}$. This indicates that there's no point of running the code over more than $1/(1-f)$ processors because any further speed-up would be spoiled by the serial portion of the code.

Unfortunately, the function $1/(1-f)$ starts growing steep only in the close vicinity of $f = 1$ (see Fig. 5.15); as an example, with 90% of the code parallelizable $(f = 0.9)$ there's no scope for more than only ten processors. To keep, say, 1000 processors busy we need applications (and algorithms) with f as high as 99.9% !

This looks like a rather gloomy limitation: fortunately, as we will see in the sequel, the fraction of CPU time spent in the parallel portion of the code is an increasing function of problem size (Gustaffson's law) so that the Amdhal's constraint mitigates as larger problems are tackled.

Depending on the type of algorithms and application, CFD is generally speaking characterized by high values of f, often beyond 0.95. Such a situation occurs whenever the serial portion of the code can be confined to the input/output activity, which is typically dwarfed by the computing section of the code.

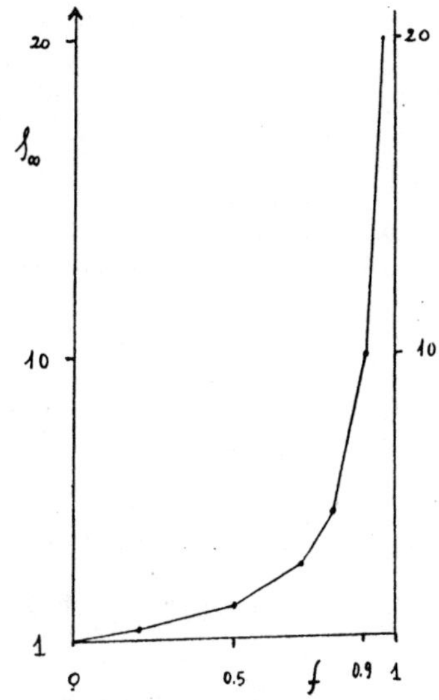

Figure 5.15: The maximum speedup S_∞ as a function of the parallel fraction f

Besides the upper bounds, it's interesting to estimate the "most reasonable" value for the number of processors to be used in a given application characterized by a fraction f of parallel content.

This information is easily highlighted by rewriting the eq. 5.8 in a self-similar form:

$$\tilde{\eta} \equiv \frac{S}{S_\infty} = \frac{1}{1 + \frac{P_C(f)}{P}} \equiv \frac{\tilde{P}}{\tilde{P} + 1} \tag{5.9}$$

where

$$S_\infty = (1 - f)^{-1} \quad \tilde{P} \equiv \frac{P_C}{P} \quad P_C(f) = \frac{f}{1 - f} \tag{5.10}$$

From these expressions we realize that the "natural" number of processors not to be overcome is P_C, i.e. the number of processors such that η achieves only half of its theoretical maximum η_∞.

For a simple mnemonics, P_C is the ratio of the fraction of parallel to serial code, and the maximum speed-up is the inverse of the fraction of serial code.

5.5.3 Load balance

Another important "caveat" for parallel computing is the issue of **load balance**. This concept is fairly straightforward: if the workload is unevenly distributed across the pool of concurrent processors, the elapsed time of the application is dictated by the slowest process. A similar conclusion applies if a homogeneus workload is distributed across a pool of heterogeneous processors.

In other words, the temporal metrics of parallel computing is the "Max" norm:

$$T_{max}(P) = \mathrm{Max}_p\{T_p\} \qquad p = 1, ..., P \tag{5.11}$$

where T_p is the elapsed time taken by the p-th processor to complete its task.

If $< T_p >$ denotes the average time over the pool of processors, load unbalancing incurs an efficiency penalty given by:

$$\eta = \frac{< T_p >}{T_{max}} \tag{5.12}$$

For a problem with, say, N units of work (for instance the previous example of summing N numbers) and P processors, with $N = P \cdot n + R$; $0 \leq R \leq P - 1$, the optimal repartition is:

R processors doing $n + 1$ units of work

P-R processors doing n units of work

The resulting load unbalance efficiency is therefore given by

$$\eta = \frac{n}{n+1} \simeq \frac{N/P}{N/P+1} \simeq 1\,;\text{for } N \gg P \tag{5.13}$$

Load unbalance becomes more and more pressing as the number of processors becomes comparable with the amount of work units to be completed, i.e. when the so called "grain" $n = N/P$ of the application becomes smaller and smaller.

Potential exposures to load unbalance are typically associated with boundary conditions, because the processors dealing with the boundaries do not generally share the same amount of work of internal processors. The most serious concerns come from CFD applications characterized by irregular and/or time-dependent data structures. Typical examples are irregular, possibly time-varying, geometrical domains and/or adaptive calculations where the density of the computational grid changes as a function of the dynamics of the physical phenomena occurring in the domain.

Dynamic load balancing is definitely one of the open frontiers of parallel CFD.

5.5.4 Coordination and communication overheads

At the bottomline of almost any parallelization strategy there is the time-honored philosophy of "divide-and-conquer": in order to solve a single, huge problem, it is often convenient to break it into a set of smaller (hence easier) problems, solve each of them separately, and finally patch the partial solutions together to produce the global one.

It is sometimes useful to visualize the divide-and-conquer strategy in terms of an analogy between processes executing on a given machine and a fictious "computational particle" representing this process [97]. Within this analogy, the "divide-and-conquer" method reads as follows

1. Particle break-up into P droplets ($t = t_1$)

2. Droplet propagation and interaction ($t_1 \leq t \leq t_2$)

3. Droplet coalescence at $t = t_2$

Here "particle" means the single (huge) problem running on a single node, while "droplet" means the corresponding sub-problem running on one of the P nodes of the parallel machine:

If $T(1)$ denotes the elapsed time required to solve the problem on a single machine (computation time), the corresponding time on the parallel machine is given by:

$$T(P) = \tau_{dis} + (\frac{T(1)}{P} + \tau_{com}) + \tau_{syn} \tag{5.14}$$

where τ_{dis}, τ_{syn} are the time-delays associated with process dispatch ("droplet breakup") and synchronization ("coalescence") respectively, and τ_{com} is the communication overhead due to the fact that in order to achieve the correct solution, the P subprocesses need to exchange some information.

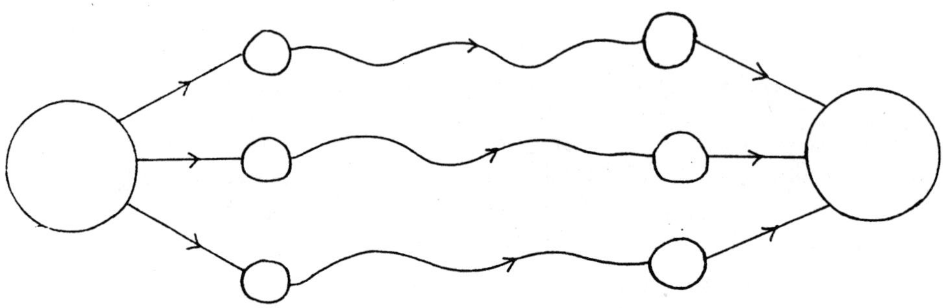

Figure 5.16: The process-particle analogy

The corresponding efficiency reads as follows:

$$\eta = \frac{\tau_{cal}}{\tau_{cal} + \tau_{dis} + \tau_{syn} + \tau_{com}} = \frac{1}{1 + \frac{\tau_{coo} + \tau_{com}}{\tau_{cal}}} \tag{5.15}$$

where $\tau_{cal} = T(1)/P$ and $\tau_{coo} = \tau_{dis} + \tau_{syn}$
is the "coordination overhead".

The loss of efficiency due to communication overheads is conveniently expressed as:

$$\eta = \frac{1}{1 + \omega} \tag{5.16}$$

where

$$\omega = (\tau_{coo} + \tau_{com})/\tau_{cal} \tag{5.17}$$

is the "Communication/Computation Ratio" (CCR). Here τ_{coo} represent coordination costs while τ_{com} represents the communication overhead; the sum of the two will be generally denoted by "communication".

The first message of eq. 5.16 is that a common goal of any parallel application is to keep ω as small as possible, i.e. the communication overhead should not exceed a negligible fraction of the computation cost. Whether or not this condition is complied with, is a question which depends on both the architectural parameters of parallel machine and the computational features of the algorithm implementing a given application.

The main architectural parameters are:

P number of processors

B interprocessors bandwidth

Λ interprocessor latency

$\mathcal{S}_1$ uniprocessor processing speed

The main application parameters are:

N global size of the problem (number of discrete degrees of freedom)

χ computational density (number of flop/s per grid point needed to advance a single computational unit of work)

σ communication density (number of bytes per grid point to be exchanged to advance a single computational unit of work)

Σ number of grid points to be exchanged by a single processor across the interprocessor boundary

As a result, the CCR becomes a function of a relatively large set of parameters. Fortunately, these parameters reorganize in such a way as to recast the functional dependence of CCR in a self-similar form as already discussed in the previous section. For the case in point, we shall consider three critical P_c's, corresponding to: communication overheads, latency overheads and synchronization overheads respectively.

Let us expand this concept in more detail. The computation cost can be expressed as:

$$\tau_{cal} = \chi n / \mathcal{S}_1 \tag{5.18}$$

where $n = N/P$ is the **local** size of the problem. Here, we shall consider the "continuum" limit $\lim P \to \infty,\ n \to 0,\ nP = const. = N$.

The communication cost takes the following form:

$$\tau_{com} = \frac{\sigma\Sigma}{B} + M\Lambda \tag{5.19}$$

where M is the number of messages required to transmit the amount $\sigma\Sigma$ of data, needed to advance one unit of computational work.

The coordination overhead can be expressed as follows:

$$\tau_{coo} = 2N_s\Lambda\phi(P) \tag{5.20}$$

where N_s is the number of synchronization points ("barriers") and $\phi(P)$ is a function depending on the way how the **global** communication is implemented. The factor 2 stems from dispatch and synchronization.

Global communication is a P^2-complex operation (each processor talking to each other) but we assume it is dealt with by a tree-algorithm bringing it down to a $\log_2 P^2$ complexity. So $\phi(P) = 2\log_2 P$.

Plugging the eq. 5.18-5.20 into the expression 5.17 for ω, we obtain the following expression:

$$\omega = aP^{\frac{1}{3}} + bP + cP\log_2 P \tag{5.21}$$

where:

$$a = \frac{\sigma}{\chi}\frac{V}{B}\frac{1}{L} \quad b = \frac{\Lambda V M}{\chi L} \quad c = \frac{2\Lambda N_s V}{\chi L^3}$$

L being the linear size of the global problem ($N = L^3$).

Note that we have made use of the equation:

$$\Sigma = L^2 P^{-2/3} \tag{5.22}$$

resulting from the fact that in the scenario where $N = const$ size of the exchange surface decreases like $P^{-2/3}$ in three-dimensions.

More generally, for a d-dimensional decomposition of a D-dimensional domain we would write

$$\Sigma = L^{D-d}l^{d-1} = L^{D-1}P^{\frac{(1-d)}{D}} \tag{5.23}$$

since $L^d = Pl^d$, l being the linear size of the local problem ($n = l^3$).

This formula highlights the superiority of three-dimensional domain decomposition in terms of minimizing the interprocessor surface. For the sake of simplicity, in the following we shall always refer to the case $D = 3$ and $d = 3$.

Also, we assume that just one layer of surface data have to be exchanged across the interprocessor boundary. This is correct only for low-order discretization schemes, while for higher-orders a "shell" of data whose width goes with the order of the discretization, needs to be exchanged. Certain algorithms may even require the exchange of volumetric data. This is case, for instance, of multidimensional Fast Fourier Transforms requiring a matrix transpose operation.

In this case Σ scales also like $1/P$, so that the notion of P_C is devoided since Σ/N is independent of the number of processors.

Each of three parameters a,b,c, serving as corresponding inverse critical numbers of processors, can be given a specific physical meaning.

Let us begin by a, and recast it into the form:

$$a = \frac{\sigma/\chi L}{B/S_1} = \frac{CCR(\text{Application})}{CCR(\text{Platform})} \tag{5.24}$$

Here the numerator can be interpreted as the CCR (Communication to Computation **R**atio) of the application, as it represents the number of Bytes to be exchanged by each processor per Flop spent into the calculation.

The denominator, on the other hand, has a similar meaning as referred to the computational platform: it represents the number of bytes which can be exchanged off-processor per Flop processed in-processor.

It becomes now clear that good performance hinges on the CCR of the platform being able to keep up with the CCR imposed by the application. Briefly, $a \ll 1$.

More precisely, we can write

$$a = \frac{L_\sigma}{L} \tag{5.25}$$

where

$$L_\sigma = \frac{\chi/\sigma}{B/S_1} \tag{5.26}$$

Here L_σ plays the role of a natural message length below which communication takes over calculation. Hence, a "good" application should feature $L \gg L_\sigma$.

As an order of magnitude consideration, a "good" parallel application should feature $\sigma/\chi \simeq 0.1$ (80 Flops per variable, i.e. 10 Flop/byte), and a "well balanced" parallel architecture $B/S_1 \simeq 0.1 Byte/Flop$.

Making abstraction of latency overheads, we write

$$\begin{aligned} \eta_\sigma(P) &= \tfrac{1}{1+x^{1/3}} \\ x &\equiv \tfrac{P}{P_{C\sigma}} \\ P_{C\sigma} &= \tfrac{L^3}{L_\sigma^3} = \tfrac{N}{N_\sigma} \end{aligned} \tag{5.27}$$

From this expression we deduce that communicativity sets a serious penalty in the "low parallelism" region $P \ll P_{C\sigma}$, while the loss of massive scalability ($P \gg P_{C\sigma}$) is rather smooth due to the slow decay of the function $(1 + x^{1/3})^{-1})$ for large x .

Some trivial, and yet in our opinion important, "caveats" are in order. It is no good to achieve high η's by raising χ because this obviously taxes the efficiency of the serial application ! Yet, in view of minimizing the overall elapsed time:

$$T(P) = \frac{\chi L^3}{P S_1 \eta(P)} \equiv \frac{T(1)}{P\eta(P)} \tag{5.28}$$

a slightly more inefficient (non minimal χ) might sometimes be preferred if its parallel efficiency can overcompensate the loss of serial efficiency.

Always in the spirit of minimizing $T(P)$ (rather than maximizing $\eta(P)$ alone !), it is unfair to disdain relatively low $\eta(P)$ due to high values of S_1 (fast processing nodes): a pool of relatively (parallel) inefficient **fast** nodes can often be more effective than a series of very efficient **slow** nodes. The message is that, when comparing algorithms and architectures, Mflop/s, Speed-up and Efficiency can be sometimes very deceptive indicators: elapsed time to solve a given problem is a more solid metric for performance evaluation.

Let us now come to the latency component of communication overheads. The physical meaning of the b parameter is readily deduced from its very expression:

$$b = \frac{M \Lambda S_1}{\chi N} \tag{5.29}$$

The numerator is the number of Flops "wasted" in latency delays, while the denominator is the number of Flops "usefully" spent to advance the computation. Thus, b represents the percentage of Flops wasted in latency versus useful Flops.

Again, we introduce the notion of a critical processor number:

$$P_{C\Lambda} = (N/N_{\Lambda}) \equiv b^{-1} \tag{5.30}$$

with

$$N_{\Lambda} = (M S_1 \Lambda / \chi) \tag{5.31}$$

Whence the self-similar expression of the efficiency (leaving aside all other delays)

$$\eta_\Lambda = \frac{1}{1+y}; \qquad y \equiv P/P_{C\Lambda} \tag{5.32}$$

In close analogy with L_σ, here $L_\Lambda = N_\Lambda^{-1/3}$ represents the "natural size" below which latency-communication effects spoil the parallel efficiency.

We note that η_Λ drops much more rapidly down than η_σ at high P: this reflects the critical role played by latency on the scalability.

In passing, we pinpoint the importance of coalescing messages ($M \simeq 1$) whenever latency is an architectural issue.

As usual, we remark the beneficial role of high computational density ($\chi \gg 1$) (See "caveat" of the previous section).

Let us now come to the global synchronization overheads. Following a familiar (by now) procedure, c can be identified as the ratio of Flops lost in synchronization delays versus "useful" Flops.

Strictly speaking, here we cannot introduce the notion of critical processor due to the scaling-breaking logarithmic term.

Notwithstanding with this "anomaly", we shall nonetheless introduce an approximate critical P_{Cs} defined as:

$$P_{Cs} \simeq N/N_s; \qquad N_s = \frac{2N_S S_1 \Lambda}{\chi} \tag{5.33}$$

after which the corresponding efficiency reads as

$$\eta_s = \frac{1}{1+z\log z}; \qquad z \equiv P/P_{Cs} \tag{5.34}$$

Here, the story goes pretty much the same way as for latency communication overheads, with N_s playing the role of M.

This points to the importance (especially on high-latency architectures) to minimize the number of synchronization points during the calculation, i.e. to work as much as possible in a **loosely synchronous** scheme. Due to the $z\log z$ dependence, the synchronization overheads are even (slightly) more stringent as long as scalability issues are concerned.

5.6 Redundancy

The best parallel algorithms are not necessarily the parallelized counterpart of the best serial ones. This stems from the fact that sometimes, in order to achieve high parallel efficiency, a number of redundant steps have to be executed which have no counterpart in the serial scheme. Clearly, this also adds to the parallel overhead.

An example of redundancy are multigrid schemes, in which the grid-coarsening procedures of the parallel scheme cannot go as deep as in the serial algorithms because the limit of one point per processor is reached when there are still P points in the computational domain. Another example, covered in this work, is domain decomposition with overlapping boundaries, in which each processor computes slightly more than strictly needed to improve convergence of parallel iterative methods.

Formally, the redundancy penalty tranlates into the following efficiency:

$$\eta_{red} = \frac{T(1)}{PT(P)}$$

where $T(1)$ refers to the **best** serial algorithm.

5.7 Summary

Summarizing, the parallel efficiency can be written as the product of four distinct sub-efficiences

$$\eta = \eta_f \cdot \eta_{lb} \cdot \eta_{com} \cdot \eta_{red} \tag{5.35}$$

corresponding to parallel "losses" due to limited fraction of parallel potential (η_f), load-unbalance (η_{lb}), communication overheads (η_{com}) and redundancy overheads (η_{red}).

These are visualized in Fig. 5-16 which represents to the time-line of a 2 process parallel scheme, a sort of "Feynman diagram" for parallel computing, as opposed to the serial time-line.

It goes without saying that, in order to achieve satisfactory efficiency, **each** of these factors must be significantly close to one. This is why picking up the enticing potential of parallel computing is a challenging task indeed.

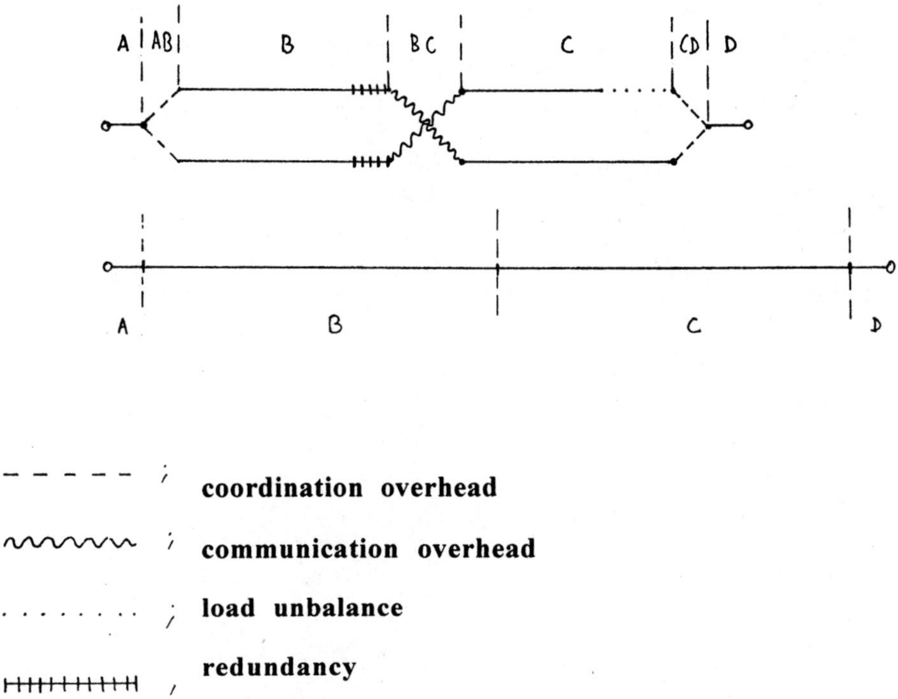

Figure 5.17: The time-line of a parallel process ($P = 2$) as compared to the serial time-line (bottom). A, B, C, D label four different stages of the calculation. Stage A and D are serial, B and C are parallelizable.

Chapter 6

Parallel Explicit Methods

Having surveyed the basic CFD algorithms in Chapters III-IV and the general aspects of parallel computing in Chapter V, we're now prepared to merge the two subjects and enter the core of this work, i.e. Parallel CFD [98] [99].

We shall begin by considering **explicit grid** methods in both local and global form. Subsequently, we shall discuss **explicit particle** methods. In Chapter VII we shall then move to **implicit grid** methods. For the sake of clarity, the above subjects are organized in two distinct chapters:

- Explicit Methods
  - Grid
    - Local (Finite Differences, Volumes, Elements)
    - Global (Spectral)
  - Particle
    - Molecular Dynamics
    - Direct Simulation Monte Carlo
    - Lattice Gas and Lattice Boltzmann
    - Vortex methods
- Implicit Grid Methods
  - Iterative
    - Relaxation
    - Gradient

– Direct

As already discussed in Chapter IV since explicit algorithms preserve causality, the local nature of the differential operators is reflected by the corresponding matrix representations on the discrete grid. Consequently, no linear algebra solvers are required ("**matrix freedom**").

Implicit algorithms, on the contrary, do violate the sequential nature of the time variable, with the result that **global** dependencies arise in the time marching procedure so that linear solvers are needed to advance the solution in time.

This basic distinction clearly indicates that explicit algorithms are much more straightforwardly parallelized than implicit ones.

Before entering specific details, let us recall that parallel opportunities come into two flavours:

- functional parallelism (at instruction, or program, level)
- data parallelism (at data level)

Since scientific problems and notably Fluid Dynamics are mostly formulated in terms of **a few** instructions ("operators") acting repeatedly on **huge** amount of data ("operands"), it is clear that data parallelism is the prime candidate for efficient parallel computing [100].

Data parallelism corresponds to concurrent processing of separate portions of a given data set, making allowance for a certain degree of intercommunication needed to recover the correct global solution.

In the case of Grid Methods, data parallelism presumes a physical decomposition of the computational domain in which the Navier-Stokes equations are to be solved: this is commonly known as **Domain Decomposition** (DD). The way to DD is by no means unique, and two broad categories can be distinguished: **geometric domain decomposition** (GDD) and **algebraic domain decomposition** (ADD) [101].

In our terminology, GDD denotes the case in which DD is achieved by up-front inspection of the geometrical/topological features of the problem **before** the discretization apparatus is put in place.

ADD, on the other hand, designates the case in which the decomposition is performed "a posteriori", on the global data structure associated with the serial problem.

GDD and ADD are by no means mutually exclusive, nor do they necessarily lead to different parallel algorithms: in general, however, they **do** differ whenever the mapping from the geometrical domain to the data structures is non trivial. Such is the case, for instance, of geometrically complex problems handled by unstructured meshes.

6.1 Parallel Explicit Grid Techniques

In this section we shall be concerned with the parallelization of explicit grid methods.

More specifically, we shall consider three major classes of algorithms:

- Local structured
- Local unstructured
- Global structured

The common thread of these otherwise fairly distinct algorithms is **matrix freedom**, i.e. the fact that since causality is preserved by the time-marching procedure, none of them requires linear algebra solvers.

Let us illustrate the basic concepts with a simple one-dimensional time dependent and compressible problem.

6.2 Structured Explicit (SE)

Let us consider the Navier-Stokes equations for a one-dimensional (weakly) compressible flow ($p \sim const.$):

$$\begin{aligned} &\partial_t \rho + \partial_x(\rho u) = 0 \\ &\partial_t u + u\partial_x u = \nu \partial_x^2 u - \tfrac{1}{\rho}\partial_x p \end{aligned} \tag{6.1}$$

($p = \rho T$) , with some initial and boundary conditions on the segment $0 \le x \le L$.

By adopting 2nd order centered-differences in space, and first-order explicit time marching, the above equations take the following discretized form (caret denoting $t + \Delta t$):

$$\begin{aligned} &\hat{\rho}_i = \rho_i - \alpha(u_{i+1}\rho_{i+1} - u_{i-1}\rho_{i-1}) \\ &\hat{u}_i = u_i - \alpha(u_i u_{i+1} - u_i u_{i-1}) + \beta(u_{i+1} - 2u_i + u_{i-1}) \\ &-\alpha(q_{i+1} - q_{i-1}) \end{aligned} \tag{6.2}$$

where $\alpha = \frac{1}{2}\Delta t/\Delta x$, $\beta = \nu \Delta t/\Delta x^2$ and the temperature T has been set to the unit value, so that $q = \ln \rho$.

In a more compact matrix notation, the above equation can be recast in the following form:

$$\begin{aligned} \hat{\rho} &= \rho - \alpha A(u)\rho \equiv \mathcal{A}\rho \\ \hat{u} &= u - \alpha C(u)u + \beta D(u) - \alpha G\rho \equiv \mathcal{B}u + \mathcal{G}q \end{aligned} \tag{6.3}$$

where the explicit form of the matrices A, C, D, G is readily derived by direct inspection of the eq. 6.2:

$$\begin{aligned} A_{ij} &= (u_{i-1}\delta_{i,j-1} - u_{j+1}\delta_{i,j+1}) \\ C_{ij} &= (u_i\delta_{i,j-1} - u_j\delta_{i,j+1}) \equiv u_j G_{ij} \\ D_{ij} &= (\delta_{i,j-1} - 2\delta_{ij} + \delta_{i,j+1}) \\ G_{ij} &= (\delta_{i,j-1} - \delta_{i,j+1}) \end{aligned} \tag{6.4}$$

The eq. 6.3 is a set of non-linear equations which must be generally dealt with by some iteration procedure.

As the simplest instance, we shall assume that the l.h.s. of eq. 6.3 is evaluated at the current iteration level, while the r.h.s. is computed at the previous one.

Thus, we're naturally led to identify each iteration cycle with the basic unit of computational work, to which parallel considerations will be applied .

Let us rewrite eq. 6.3 in a compact form as:

$$\begin{aligned} \hat{\rho}^{k+1} &= \mathcal{A}(u^k)\rho^k \\ \hat{u}^{k+1} &= \mathcal{B}(u^k)u^k + \mathcal{G}q^k \end{aligned} \tag{6.5}$$

where k refers to the iteration cycle.

From this expression, we observe that the computational kernel of our explicit algorithm is constituted by three (sparse) matrix-vector scalar products. Thus the problem of parallelizing the eq. 6.1 is reconduced to the task of parallelizing a basic linear algebra operation such as sparse matrix-vector-multiply (**mvm**).

In particular, since for our one-dimensional example the matrices $\mathcal{A}, \mathcal{B}, \mathcal{G}$ have a tridiagonal form, we shall focus our attention to the problem of parallelizing the matrix-vector product:

$$y = Tx \tag{6.6}$$

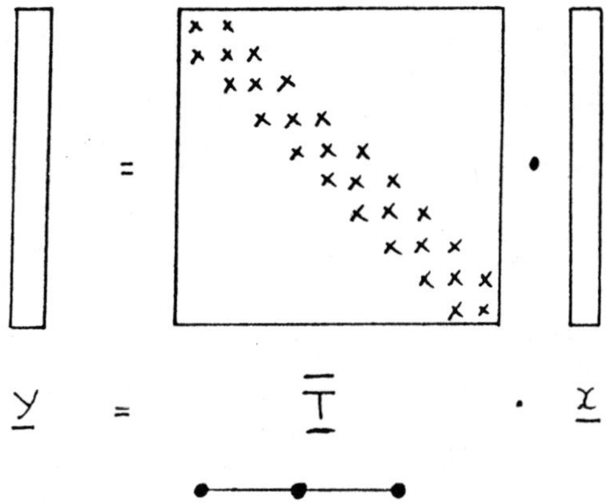

Figure 6.1: Matrix-vector multiply $y = Tx$ with T tridiagonal

where T is a generic tridiagonal matrix of size $N \times N$

$$T_{lm} = a\delta_{l,m+1} + b\delta_{l,m} + c\delta_{l,m-1} \;, l, m = 1, N \tag{6.7}$$

Suppose for simplicity we are given just two processors ($P = 2$). By splitting the source (x) and target (y) arrays in two independent subvectors of size $N/2$: $x = x^1 + x^2, y = y^1 + y^2$, the scalar product 6.6 yields:

$$y^1 + y^2 = T^{11}x^1 + T^{12}x^2 + T^{21}x^1 + T^{22}x^2 \tag{6.8}$$

where T^{pq} is the matrix partition induced by the decomposition of the arrays x and y. Since the subvectors are independent, the expression 6.8 splits into two independent relations:

$$\begin{aligned} y^1 &= T^{11}x^1 + T^{12}x^2 \\ y^2 &= T^{21}x^1 + T^{22}x^2 \end{aligned} \tag{6.9}$$

Assuming processor 1 owning T^{11}, T^{12}, x^1 and processor 2, T^{21}, T^{22}, x^2, the operations 6.9 can proceed concurrently except for the cross-diagonal terms $T^{12}x^2$ and $T^{21}x^1$ which imply some degree of interprocessor communication.

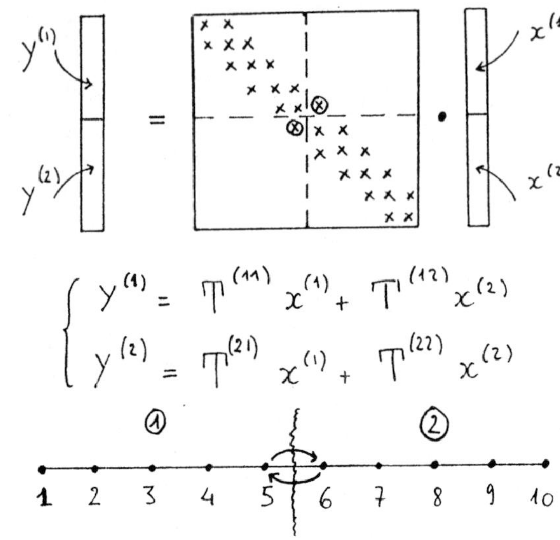

Figure 6.2: Matrix partitioning and the corresponding domain decomposition associated with the partition 6.10

For the sake of clarity let us hereafter refer to a specific partition corresponding to cutting the arrays in two as follows (see Fig. 6.2):

$$\begin{array}{lll} x^1_{i_1} = x_i & i = 1, N/2; i_1 = 1, N/2 \\ x^2_{i_2} = x_i & i = N/2+1, N; i_2 = i, N/2 \end{array} \tag{6.10}$$

The corresponding matrix partition is illustrated in Fig. 6-2

This simple algebric domain decomposition has a straightforward **geometric** domain decomposition counterpart: the computational domain is broken into two compact non-overlapping segments as depicted in Fig. 6.2.

Since the nodes of this domain are connected by a 3-point stencil (represented by the tridiagonal matrix T), the border nodes need some information, [1] from the other domain in order to be correctly updated (i.e. to produce the correct values of y).

Minimization of this amount of information is therefore a primary issue for any parallelization procedure dealing with distributed memory platforms. As anticipated in the previous section, the identification of the "best" partitioning for a simple and structured geometry is a trivial task; on the other

[1]The portion of x^1, x^2 acted upon by T^{21}, T^{22} respectively

hand, for arbitrarily complex geometries the problem of identifying a minimal surface/volume partition with (almost) equal computational work across all processors becomes NP-hard.

At this stage it is of some interest to get a feel of the way that the parallel tridiagonal matrix-vector multiply would be implemented on the two main classes of parallel architectures: distributed memory MIMD, and distributed memory SIMD

6.2.1 Distributed memory MIMD

In this case, each processor owns its subportion of the array x, say $x1, x2$ for processor 1 and 2 respectively, and the corresponding submatrices T^{11}, T^{12} and T^{21}, T^{22}. As a result, interprocessor communication must take place via explicit exchange of messages between the two processors.

A parallel pseudo-program may look as follows:

```
Processor 1                          Processor 2

Send  X1(N/2) to proc. 2             Receive X1(N/2) from proc. 1
                                             and store in  X2(0)

Receive X2(1) from proc. 2           Send X1(1) to proc. 1
    and store in X1(N/2+1)

For i=1,N/2                          For i=1,N/2
 y1(i)=a*X1(i-1)+b*X1(i)+             y1(i)=a*X2(i-1)+b*X2(i)+
       c*X1(i+1)                            c*X2(i+1)
```

A few considerations are in order. First, we note that each processor has its own set of variables, matrices and a corresponding **numbering** scheme.

A complete command over this information is needed to correctly put the **explicit message-passing** (EMP) code in place (Send/Receive primitives).

This is a blame and a blessing at a time: it's a blame because the burden of correct message-passing is entirely on programmer responsibility. This results in a relatively high exposure to manual programming errors as easy to generate as hard to debug.

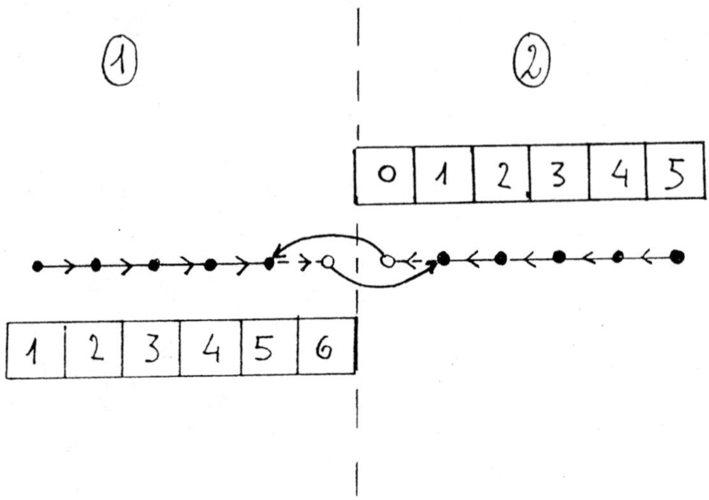

Figure 6.3: Halo cells for interprocessor communication

On the other hand, it is also a blessing because, by forcing the programmer to a in-depth comprehension of the communication pattern, it allows him to achieve the highest efficiency.

Finally, note the presence of "halo" cells (numbered 6 for processor 1 and 0 for processor 2 in Fig 6.3), which serve as temporary buffers for the interprocessor communication.

A more detailed message passing (pseudo)-code would look as follows:

1. query(iproc, nprocs)
2. if (iproc=0) then
3. send($x(N/2)$, 4, iproc+1, 100*iproc))
4. receive($x(N/2+1)$, 4, iproc+1, 100*(iproc+1)+1)
5. else
6. receive($x(0)$, 4, iproc-1, 100*(iproc-1))
7. send($x(1)$, 4, iproc-1, 100*iproc+1))
8. endif

9. do i=1,N/nprocs

10. $y(i) = a * x(i-1) + b * x(i) + c * x(i+1)$

11. end do

A detailed comment is in order.

Statement 1: each processor must know its identity (normally called task-identifier) within the pool of participating processors. This is a number in the range [0, P-1] and is normally obtained via a query call to the system.

Statement 2: Once each processor knows its identity, it can poise itself to different actions (MIMD nature of the machine is apparent here) as detailed below.

Statement 3: In our case, if iproc = 0 (left-processor), it sends the right-boundary value $x(N/2)$ (first argument of the send statement), consisting of 4 bytes (2nd argument) to its right neighbor processor, identified by the integer iproc+1 (3rd argument).

Subsequently (statement 4) it receives the new right-boundary value from processor 1 (right neighbor).

Note that each message gets an integer identifier (4th argument). In our example we stipulate the (arbitrary) convention that the identifier is *100*iproc* if the message is sent rightwards by processor *iproc*, and *100*iproc +1* if it is sent leftwards.

If the processor id is 1, the same procedure is executed in reverse order: the new left-boundary value $x(0)$ is received from processor 0 to which the actual left-boundary value $x(1)$ is forwarded (statements 6 and 7).

Once the message-passing operation is completed, each processor can proceed independently upon its share of data (statement 9-11) like in a serial code.

The element $x(N/2)$ in the left-processor 0 is received as $x(0)$ by the right-processor 1, and, conversely, the element sent by processor 1 as $x(1)$ is received by processor 0 as $x(N/2+1)$.

This example, albeit oversimple, shows how explicit message-passing forces the programmer to keep full command at the details of the code. It also shows the explicit coding overhead: statements 1 to 8 serve as a prologue to the actual calculation which only takes 3 statements (9 to 11). This is often perceived as a significant burden as well as potential source of programming errors, and motivates the conspicuous efforts currently devoted to lay down message-passing programming standards.

6.2.2 Distributed memory SIMD

Regular, structured matrix-vector multiply a typical prototype of a SIMD operation. No surprise, therefore, that SIMD machines cater for specialized "stencil communication software" dealing with inter-processor communication in a way which is completely transparent to the programmer [102].

A typically "array-language" like program on a SIMD machine may look as follows:

$$y = a * f\_\text{grad}(x) + (b - a - c) * x + c * b\_\text{grad}(x) \tag{6.11}$$

Here, a, b, c are three N-dimensional arrays representing the tridiagonal matrix T_{lm}, f_grad and b_grad are typical "stencil-software" operators yielding the backward and forward gradient of the vector x_i respectively:

$$\begin{aligned} f\_\text{grad}(x) &:= x_{i+1} - x_i \\ b\_\text{grad}(x) &:= x_i - x_{i-1} \end{aligned} \tag{6.12}$$

The elegance and compactness of such a "code" is hard to escape and to beat. All the details pertaining to the interprocessor data exchange are left with the compiler, thus relieving the programmer from the burden of specifying and coding knotty communication statements. Unfortunately, however, such an elegant code is easily written only in the presence of **regular and simple** data structures, such as those generated by structured problems.

More general and complex problems can also be tackled, but special tricks and coding subtleties are required, which in our opinion expose SIMD programming to the same kind of error-proneness already mentioned for Explicit Message Passing.

Coming up with a program as simple as 6.11 for an irregular problem is an ambitious goal which involves a lot of computer-science research, especially in the area of parallel compilers.

Although massive efforts are being spent in this direction it seems like the advent of a reliable and general autoparallelizing compiler for distributed memory machine is still a relatively long-term target [103].

6.2.3 Efficiency of structured explicit schemes

As already noted, structured explicit schemes are typical prototypes of SIMD operation and consequently ideally suited to SIMD architectures with mesh configuration.

With the recent advances of networking and communication technology, distributed memory MIMD machines can also offer competitive performance on SIMD problems.

Just as a qualitative reference frame, let us estimate the parallel performance of a three-dimensional sparse matrix-vector-multiply on the following target distributed memory MIMD machine:

$$\mathcal{S}_1 = 100Mflop/s,\; B = 10MB/s,\; \Lambda = 100\mu s$$

Let us begin with sheer communication overheads.

For an epta-diagonal matrix multiply, a simple operation count yields $\chi = 13$ flop/s per grid-point and $\sigma = 1$ word $= 8$ bytes/grid-point.

According to the discussion of Chapter V (see eq. 5.26), we have

$$L_\sigma = 4d\frac{\sigma/\chi}{B/\mathcal{S}_1} \simeq 80d/3 \tag{6.13}$$

Where we remind that d is the dimensionality of the partitioning scheme and the factor four, instead of two, stems from bidirectional flow (send/receive) across interprocessor boundaries.

The corresponding efficiency reads:

$$\eta_\sigma = \frac{1}{1 + P^{1/d}(L_\sigma/L)} \tag{6.14}$$

Let us now come to latency overheads.

Counting $M = 2d$ messages per processor, the formula 5.31 yields:

$$N_\Lambda = \frac{M\mathcal{S}_1\Lambda}{\chi} \simeq 5000d/3 \tag{6.15}$$

and a corresponding efficiency:

$$\eta_\Lambda = \frac{1}{1 + P\frac{N_\Lambda}{N}} \tag{6.16}$$

Finally, synchronization overheads.

With just one synch point per iteration, we obtain:

$$\begin{aligned} N_s &= \frac{2N_s\mathcal{S}_1\Lambda}{\chi} \simeq 3000 \\ \eta_s &= \frac{1}{1+P\log_2 P(\frac{3000}{L^3})} \end{aligned} \tag{6.17}$$

Take now the case of three-dimensional partitioning ($d = 3$) of a cube of length $L = 100$.

The corresponding critical processor numbers are given by ($N \equiv L^3$), ($N_\sigma \equiv L_\sigma^3$):

$$P_{C\sigma} = \frac{N}{N_\sigma} \simeq 2$$

$$P_{C\Lambda} = \frac{N}{N_\Lambda} \simeq 200 \tag{6.18}$$

$$P_{Cs} = \frac{N}{N_s} \simeq 330$$

Manifestly, communication overheads act as the worst parallel performance inhibitor. The main cause is the poor computational density (less than 2 bytes per flop), a typical "drawback" of explicit algorithms.

Due to the sharp (cubic) dependence of $P_{c\sigma}$ upon χ, a relatively moderate increase of compute density is sufficient to lift this limitation. For instance, $\chi = 50$ would already result in $P_{c\sigma} \simeq 128$.

It must be emphasized that these figures should not be taken for more than a qualitative guidance, the main message being that computational leanness is a potentially serious performance inhibitor for SE schemes.

SE schemes are the simplest instance of parallel CFD; yet they prove quite useful for basic studies of transitional flows [104] [105], and it is fair to say that the advent of parallel computers has injected renewed interest into explicit CFD schemes.

6.3 Unstructured explicit (UE)

As pointed out in Chap IV, structured meshes are often too restrictive for geometrically demanding problems [106].

Geometric flexibility can be achieved by resorting to fully **unstructured** meshes in which each lattice node is generally surrounded by an arbitrary number of neighbors. Typical shapes of the corresponding finite elements, or volumes, are triangles in two-dimensions and tetrahedra in three.

The lack of structure in the mesh produces two major effects on the structure of the matrices arising from the discretization of the differential operators:

a The number of non-zero entries per row is not a constant

b The sparsity pattern of the matrices has no regularity

Point a) is a direct consequence of the fact that the lattice coordination number (i.e. number of neighbors of a given node) can vary from node to node.

Point b) is strictly related to the **mesh numbering** scheme.

It is clear that, in the absence of a regular structure, there is no obvious, natural, way of ordering the nodes of the lattice.

This puts a high incentive on the search of optimal numbering schemes devised in such a way as to generate the "best" matrices in terms of the parameters which control the computational efficiency [107].

The importance of a well-designed numbering scheme is a long-standing subject of research in the area of direct solvers, let alone parallel computing: the key point is to minimize the matrix bandwidth, which is the most demanding parameter for the computational efficiency.

In the case of parallel computing, new quantities arise which need be optimized to preserve good efficiency. These are:

a Surface/Volume ratio (Perimeter)

b Load Balancing (both computation and communication)

In order to discuss these two aspects in some detail, let us refer to the basic kernel of an UE scheme, i.e. the matrix-vector scalar product:

$$y_i = \sum_{j \in \eta_i} A_{ij} x_j \tag{6.19}$$

where A_{ij} is a sparse matrix with arbitrary sparsity pattern.

Here η_i denotes the neighborhood of node i. In a structured application η_i is identified by a triplet of integer displacements $\Delta_x, \Delta_y, \Delta_z$:

$$\eta_i = \{i \pm \Delta_x, j \pm \Delta_y, k \pm \Delta_z\} \tag{6.20}$$

which reflect the regularity of the corresponding matrix sparsity pattern (see Fig. 6.1). In the case of an unstructured application, η_i has no *a priori* regularity and consequently needs to be specified by a list of neighbors

$$\eta_i = \{j_1^i, j_2^i, \ldots, j_{\nu_i}^i\} \tag{6.21}$$

where η_i is the number of neighbors of node i.

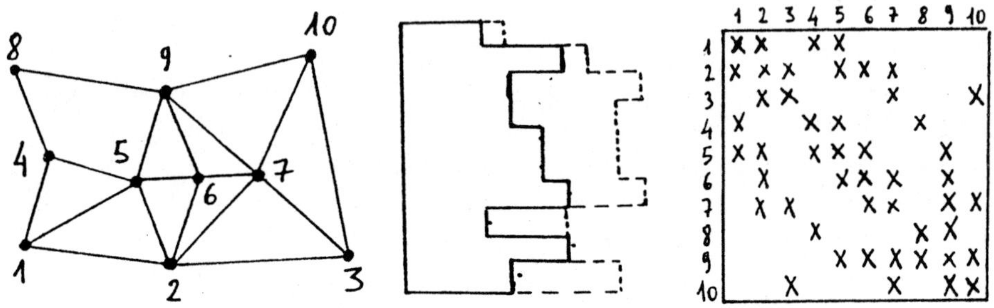

Figure 6.4: An unstructured computational grid and the corresponding matrix. Besides the matrix, the number of non-zero entries η_i (solid line) and the local bandwidth B_i (dashed line) are also shown

It is clear that if η_i is a wild function of i, it is not obvious how to partition the matrix T in such a way as to evenly distribute the workload across the processors, while minimizing at the same time the amount of interprocessor communication (see Fig. 6.4).

Formally, let $\mathcal{I}_p$ denote the set of nodes belonging to processor p under a given partition $\mathcal{P}$:

$$\mathcal{I}_p = \{i_1, i_2, ..., i_{N_p}\} \qquad p = 1, P \tag{6.22}$$

Also let T^{pp}, T^{pq} the corresponding on-block and off-block submatrices of matrix T. On the assumption that each grid point takes the same amount of compute work, and made one this amount, the workload W is defined as (we assume a constant amount of work per grid-point, as befits an explicit method):

$$W = \sum_{i=1}^{N} \nu_i = \sum_{p=1}^{P} \sum_{i \in \mathcal{I}_P} \nu_i = \sum_{p=1}^{P} \nu_p \tag{6.23}$$

The amount of interprocessor communication between processors p and

q is given by:

$$\mathcal{C}_{pq} = \sum_{i \in \mathcal{I}_P, j \in \mathcal{I}_q} a_{ij} \tag{6.24}$$

where a_{ij} is the **adiacency matrix** associated with T, i.e.

$$a_{ij} = \begin{cases} 1 & \text{if } A_{ij} \neq 0 \\ 0 & \text{otherwise} \end{cases} \tag{6.25}$$

Manifestly, with this definition $\mathcal{C}_{pq}$ represents the number of non-zero entries in the sub-block T_{pq}, and consequently

$$W_p = \sum_{p=1,P} \mathcal{C}_{pp}$$

is proportional to the the amount of work performed by processor p, whereas

$$C_p = \sum_{q \neq p} \mathcal{C}_{pq}$$

is proportional to the amount of communication involving processor p. Points a) and b) lead to the following optimization problem:

"Given a set of nodes

$$\mathcal{I} = \{x_1, x_2 ... x_N\};$$

find a partition $\mathcal{P}$

$$\mathcal{I} = \mathcal{I}_1 \bigcup \mathcal{I}_2 .. \bigcup .. \mathcal{I}_P,$$

such that:

$q_1 := Max_{p,q}|W_p - W_q|$ (computational load unbalance)

$q_2 := Max_{p,q}|C_p - C_q|$ (communication load unbalance),

and

$q_3 := Max_p C_p$ (interprocessor communication)

are minimized", under the constraint:

$$q_4 := \sum_p \{\mathcal{C}_{pp} + \sum_{q\neq p} \mathcal{C}_{pq}\} = const.$$

(number of connected nodes in the global network).

For a general geometry, this optimization problem is NP-hard, i.e. cannot be solved exactly within a polynomial number of steps.

As a result, the current trend is towards powerful heuristics yielding "quasi-optimal" solutions in a reasonably short time.

Let us clarify this matter with some illustrative examples.

Consider a 10-node mesh, numbered in x-y lexicographic order under the following partition (see Fig. 6.5):

$$\mathcal{P}^{(1)} : \mathcal{I}_1 = \{1, 2, 3, 4, 5\};\ \mathcal{I}_2 = \{6, 7, 8, 9, 10\} \tag{6.26}$$

Assuming the matrix A_{ij} has exactly the same structure as the grid connectivity, visual inspection of the matrix A_{ij} yields:

$$\mathcal{C}_{11} = 17\ , \mathcal{C}_{22} = 17\ , \mathcal{C}_{21} = \mathcal{C}_{12} = 7 \qquad (q_4 = 48)$$

Thus, $\mathcal{P}^{(1)}$ is perfectly load balanced ($\mathcal{C}_{11} = \mathcal{C}_{12}$) with a surface/volume ratio of 7/17.

Consider now a second partition $\mathcal{P}^{(2)}$:

$$\mathcal{P}^{(2)} : \mathcal{I}_1 = \{1, 3, 4, 8, 10\};\ \mathcal{I}_2 = \{2, 5, 6, 7, 9\} \tag{6.27}$$

A direct count yields:

$$\mathcal{C}_{11} = 11\ , \mathcal{C}_{22} = 21\ , \mathcal{C}_{21} = \mathcal{C}_{12} = 8 \qquad (q_4 = 48)$$

Thus, $\mathcal{P}^{(2)}$ is heavily unbalanced ($(\mathcal{C}_{22} - \mathcal{C}_{11})/\mathcal{C}_{11} = 10/11$) and features a high surface/volume ratio: 8/11 and 8/21 for processor 1 and 2 respectively.

Although both partitions are made up of the same number of nodes, $\mathcal{P}^{(1)}$ is clearly much higher quality than $\mathcal{P}^{(2)}$.

As already pointed out, finding the best partitioning is a NP-hard problem, which is why in practice one resorts to heuristics.

Among others, an important leading criterion relates to the average matrix bandwidth, $\bar{B}$, defined as the average logical distance of the matrix elements:

$$\bar{B} = \frac{1}{N} \sum_{i=1}^{N} B_i$$

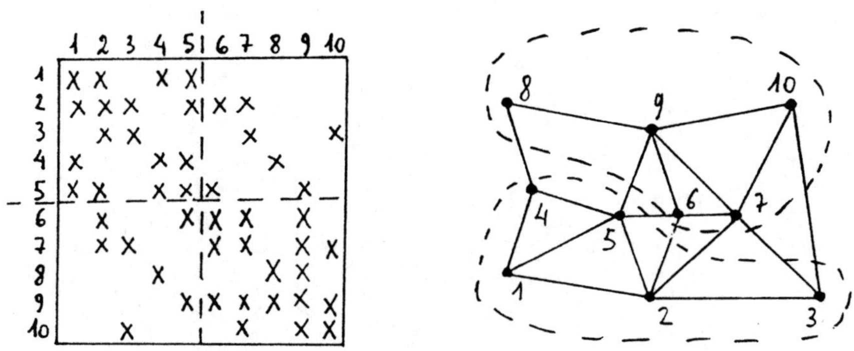

Figure 6.5: An unstructured grid partitioned according to $\mathcal{P}^{(1)}$ and the corresponding partitioning induced on the matrix A

This is compared to the average matrix line density:

$$\bar{\nu} = \frac{1}{N}\sum_{i=1}^{N} \nu_i$$

to yield $\bar{B}/\bar{\nu} \geq 1$ as a natural measure of unstructuredess.

A direct count on the "sky-line" of matrices $A_{ij}^{(1)}$ and $A_{ij}^{(2)}$ yields $\bar{B}^{(1)} = 7.5$ and $\bar{B}^{(2)} = 8.9$, i.e. $\bar{B}^{(1)}/\bar{\nu} = 1.25$ and $\bar{B}^{(2)}/\bar{\nu} = 1.78$.

Albeit trivial, this example highlights the importance of node ordering not only for serial performance, but also for load-balance and interprocessor communication issues.

As a further example take the same 10-node grid numbered this time in yx rather than xy lexicographic order.

Under the partition:

$$\mathcal{P}^{(3)} : \mathcal{I}_1 = \{1, 2, 3, 4, 5\};\ \mathcal{I}_2 = \{6, 7, 8, 9, 10\} \tag{6.28}$$

we obtain

$$\mathcal{C}_{11} = 17\ , \mathcal{C}_{22} = 19,\ \mathcal{C}_{12} = \mathcal{C}_{21} = 6\ (q_4 = 48)$$

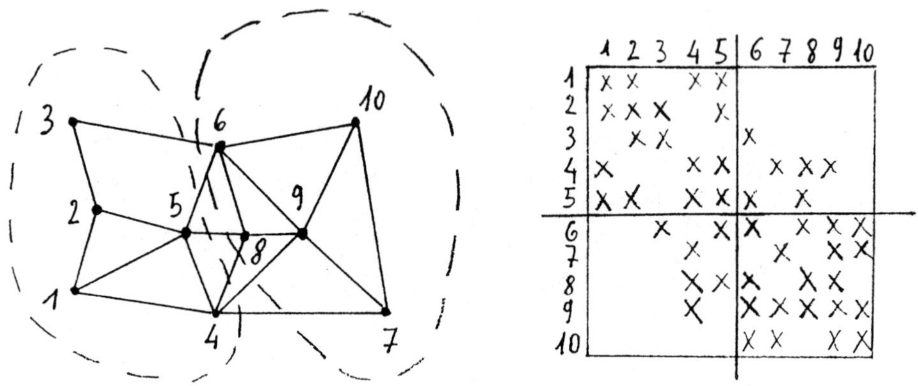

Figure 6.6: The same unstructured grid of Figg. 6.4- 6.5 numbered in y-x lexicographic order, and the corresponding matrix A

This partition features $\bar{B}^{(3)}/\bar{\nu}^{(3)} = 5.8$, which is manifestly better than 6.26 and 6.27.

Interprocessor communication can be further decreased by "moving" node 5 from $\mathcal{I}_1$ to $\mathcal{I}_2$; the new partition would achieve $\mathcal{C}_{12} = \mathcal{C}_{21} = 5$, i.e. a minor reduction of surface/volume, but only at the expense of a severe load-unbalance $\mathcal{C}_{11} = 11, \mathcal{C}_{22} = 25$.

This is because node 5 is very "heavy", in that it carries a high connectivity (5 neighbors).

This strong sensitivity to minor perturbations is a distinct manifestation of the NP-complexity of the optimization problem.

Let us now come back to the general optimization problem previously introduced, and briefly comment on the heuristic methods which are used to (approximately) solve it.

All of these methods start with the formal representation of an unstructured mesh as an **undirected graph** $\mathcal{G} = \{V, E\}$, consisting of a set of nodes, or **vertices** V, and a set of links, or **edges** E, emanating from the vertices.

In terms of graph theory the partitioning problem is to find a set of disjoint vertex subset $\{V_p\}$ such that the difference in size of the various $\{V_p\}$ is minimal, as well as the size of the **cut sets of edges** $\sum_{p,q} |C_{p,q}|$, where $C_{p,q} = \{(v, w), v \in V_p, w \in V_q\}$ and $|C_{p,q}|$ is the number of cut edges between V_p and V_q.

Let us briefly list the main graph-partitioning algorithms which have proven useful in the recent years [108].

The Greedy algorithm (GR)

This algorithm proceeds by "biting" into the mesh (whence the name "Greedy") to construct the subdomains. Since only mesh connectivity is required as an input, GR is one of the fastest graph-partitioning schemes available to date.

Manifestly, GR is a geometric domain decomposition based scheme.

The Reverse Cuthill-McKee algorithm (RCM)

This is an algebraic domain decomposition algorithm. The idea is that if an equal number of consecutive columns of a $N \times N$ banded matrix are assigned to each of the P available processors, then the maximum number of processors which need to communicate is given by $P \times B_{max}/N$, where B_{max} is the maximum bandwidth.

Therefore, this method focuses on bandwidth minimization by means of standard bandwidth reduction schemes.

RCM is very robust, but offers no guarantee of optimal surface/volume ratios, and therefore it may not be suitable for parallel processors with low-latency and low bandwidth.

The Principal Inertia algorithm (PI)

This scheme is based on the intuitive idea of slicing an object. The idea is to define a direction ξ and then project the mesh nodes onto it.

The projected points are then sorted along ξ and gathered into P sequential subsets. A reasonable choice for ξ is any of three principal inertial directions.

The PI can also be implemented in recursive form (RPI): at each recursive step, the principal inertial directions of the current mesh partition must be found.

Recursive Graph Bisection (RGB)

This is a close relative of RPI, with the basic distinction that, rather than euclidean metric, a graph metric is used instead. As a result, RGB exploits connectivity information whereas RPI does not.

Recursive Spectral Bisection (RSB)

This is probably the most abstract and most elegant partitioning scheme on the "market". It is also the one that has captured most of the attention of the parallel computing community.

The method builds upon a RGB strategy based on the idea of computing the eigenvector $\underline{x}_2$, corresponding to the second smallest eigenvalue of the Laplacian matrix $\mathcal{L}(\mathcal{G})$ of the graph $\mathcal{G}$, defined as:

$$\mathcal{L}_{ij} = \begin{cases} -1 & \text{if} a_{ij} \neq 0 \\ \nu_i & \text{if } i = j \\ 0 & \text{otherwise} \end{cases}$$

Here a_{ij} the adjacency matrix defined previously and ν_i is the degree of the i-th node in the graph (number of lines that emanate from this node).

The eigenvector $\underline{x}_2$, also known as Fiedler vector, contains important directional information about the graph: the components of $\underline{x}_2$ are weights on the corresponding vertices of $\mathcal{G}$, so that components differences provide effective information about distances between the vertices.

Sorting the vertices by the components of $\underline{x}_2$ is therefore "de-facto" a method to order them in such a way as to put neighboring vertices close one to another.

RSB is in principle quite expensive -like most eigenvalues solvers- but recent developments based on multigrid variants of the Lanczos algorithms have brought it down to acceptable levels of performance even on simple workstations (few tens of minutes on meshes with several thousands nodes).

6.3.1 Parallelization of Matrix-Vector-Multiply

Having discussed in some depth the issue of load-balancing, let us now consider the parallelization of the matrix-vector-multiply (*mvm*):

$$y_i = \sum_{j \in \eta_i} A_{ij} x_j$$

on a distributed pool of P processors.

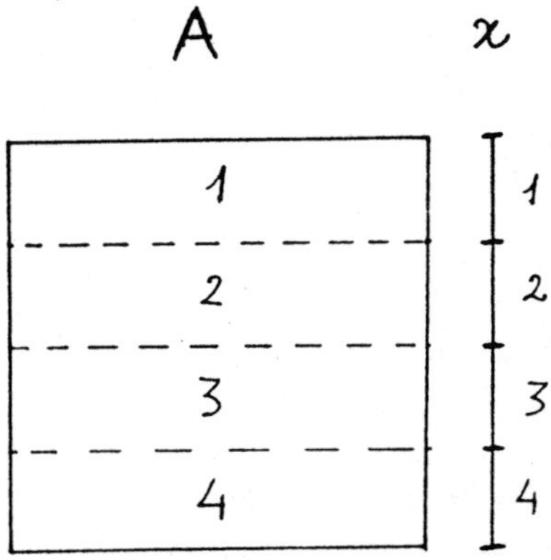

Figure 6.7: A one-dimensional "striped" partitioning of matrix A and source vector x, with $P = 4$

Here η_i denotes the neighborhood of the i-th node described by a list of nodes.

The sparsity pattern of the matrix A is characterized by two main parameters:

local bandwidth B_i: maximal logical distance between non-zero elements in the i-th row

ν_i; number of non-zero entries in the i-th row

We shall assume that a suitable node ordering has been adopted so as to minimize load unbalance.

Let us consider a fully distributed implementation, in which each processor owns a horizontal strip of A_{ij}, $A_{ij}^{(p)}$, and the corresponding segment $x^{(p)}$ of the source vector x (see Fig. 6.7)

The distributed *mvm* looks as follows (summation over j is implied)

$$y_i^{(p)} = A_{ij}^{(pp)} x_j^{(p)} + \sum_{q \neq p} A_{ij}^{(pq)} x_j^{(q)} \quad p = 0, 1, ..., P-1$$

The first term in the r.h.s. corresponds to the "intra-processor" work, while the off-diagonal terms require the corresponding portion of x being transmitted from processor q to processor p.

The cost of this operation depends on

i) Amount of data to be transmitted (roughly the longest segment subtended by the matrix profile within the off-diagonal block $A^{(pq)}$.

ii) The interprocessor distance $|p-q|$

For most modern architectures, in which the interconnecting network is designed in such a way as to make all processors equidistant (metric-free), the item ii) can be safely neglected.

Let us now consider the extreme case of "shared-memory-like" implementation. Here, besides the horizontal strip $A^{(p)}$ each processor owns a full copy of the source vector x; i.e. $x^{(p)} = x$.

Under these conditions, interprocessor communication is clearly brought down to zero. The price, however, is breaking of scalability at the level of memory requirements. Assuming in fact a storage format whereby only non-zero entries of A are stored, the total memory requirements are:

$$M = (\nu + 1)N$$

where ν is the average number of non-zero entries $\nu = \sum_{i=1}^{N} \nu_i / N$ and the extra factor one comes from the vector x.

In a scalable version the memory allocated by each processor, M_p, should scale like $1/P$.

Here, however, we have:

$$M_p = (\frac{\nu}{P} + 1)N; \ M \equiv \sum_p M_p = \nu N + PN$$

which only scales for $P \ll \nu$, i.e. virtually zero to any practical purpose.

The "quasi-shared-memory" approach is consequently restricted to a small number of processors.

Let us now consider an intermediate version in which, besides A^p, each processor owns a "redundant" segment of the vector x, where redundant means that there is a certain degree of overlapping δ_p, between the various segments x_p:

$$\delta_p = |x_p \bigcap x_{p+1}|/N; \quad (0 \le \delta_p \le 1)$$

Here $|...|$ means the number of elements.

Manifestly, $\delta_p = 0, 1$ correspond to fully distributed and "quasi-shared-memory" implementation respectively.

Roughly speaking, this scheme achieves a reduction of communication cost by a factor $(1 - \delta)$, δ denoting a suitable average over all processors, while computation and storage requirements increase by a corresponding factor $(1 + \delta)$ per processor.

Depending on the architectural parameters, one may want to play with the overlap parameter and sacrifice some CPU power and memory words to reduce the communication overhead.

6.3.2 Efficiency of Unstructured Explicit schemes

Unstructured explicit schemes are subject to the same performance inhibitors as structured ones, notably poor compute density. In addition, they are exposed to potentially severe limitations imposed by load unbalance, both in computation and communication. Lack of structure may, however, also result in some marginally beneficial effect. Due to random access to in and off-processor memory, the effective speed S_1 and the effective bandwidth B of an UE scheme generally undergoes a degradation with respect to the corresponding SE scheme.

Should the former degradation be more pronounced than the latter, an improved parallel efficiency would result. This should not hailed as a great achievement in itself since it is clear that overall performance is what counts in practice. A second beneficial effect may be observed at the level of global synchronization, when a pool of processors (slaves) need to send a message to a single (master) processor.

In a structured, load-balanced, application the requests to send issued by the various slaves are very likely to occur virtually at the same time, thus generating a burst of communication which may result in anomalous "wait-status" conditions very detrimental to the overall efficiency. Such coherent bursts of communication are obviously much less likely to occurr in (slightly) load-unbalanced applications.

Although this sort of "collective effects" escape a clean theoretical analysis, they may well play a significant role in real life applications.

6.4 Parallel Spectral Methods

Spectral Methods (SM) have been the workhorse of turbulence calculations for more than 20 years. Their success draws upon the clever Fast-Fourier

Transform (FFT) algorithm, which manages to turn a $O(N^2)$ complex problem into a $O(N \log_2 N)$ complex one.

At a first glance SM do **not** stand out as prominent candidates for parallel computing.

To see why, let us recall the very definition of the discrete Fourier Transform (DFT) of a periodic function $f(x)$ of period L:

$$\hat{f}(k_l) = \Delta \sum_{j=0}^{N-1} e^{-ik_l x_j} f(x_j) \equiv \sum_{j=0}^{N-1} E_{lj} f_j$$

$$x_j = \Delta j; \quad k_l = \frac{2\pi}{L} l; \quad l = 0, N-1 \tag{6.29}$$

where $L = N\Delta$, N being the number of discrete points spaced Δ apart.

This definition highlights the **global** nature of the DFT, hence its highly communicative character: evaluation of a DFT at a single point in k-space necessitates knowledge of the entire set of values in x-space.

Despite its high degree of communicativity, a number of DFT algorithms conducive to parallel computing have been worked out in the last decade [109].

Generally speaking, these methods are based upon simple tensorial decomposition of the cubic computational domains as a direct product of lower dimensional cubes (planes in $D = 2$ and lines in $D = 1$).

Denoting by X a 3D cube, a possible decomposition reads:

$$X = \Pi_{xy} \otimes \Lambda_z$$

where Π_{xy} is a plane at z=const. and Λ_z is the line x=const., y=const. (Fig. 6.8)

Let us now consider the problem of distributing a three-dimensional FFT across a set of P processors, each owning a slice of data $u(j_1, j_2, j_3)$ associated with the plane j_3=const.$=0, 1 \ldots N-1 = P-1$.

As a first step, each processor can FFT its own slice of data in a completely independent fashion:

$$\hat{u}(l_1, l_2, j_3) = FFT_{xy}\{u(j_1, j_2, j_3)\}; \; j_3 = 0, 1, \ldots, N-1$$

To proceed further, however, each processor needs to import a plane of data, say $\hat{u}(l_1, l_2, j_3 = 0, N-1)$ from every other processor (see Fig. 6.9).

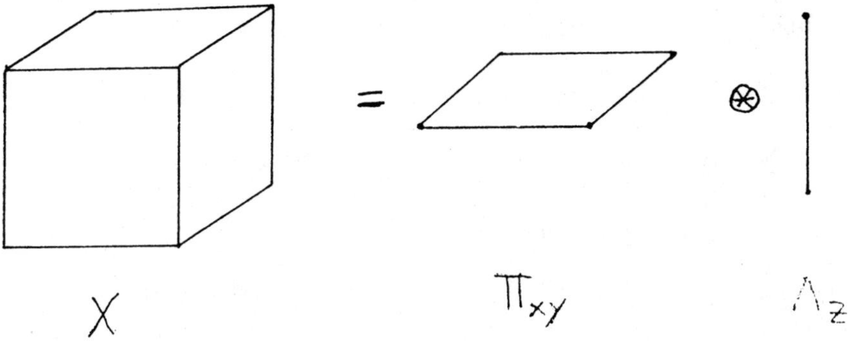

Figure 6.8: The cube X as a direct product of the plane Π_{xy} and the pencil Λ_z

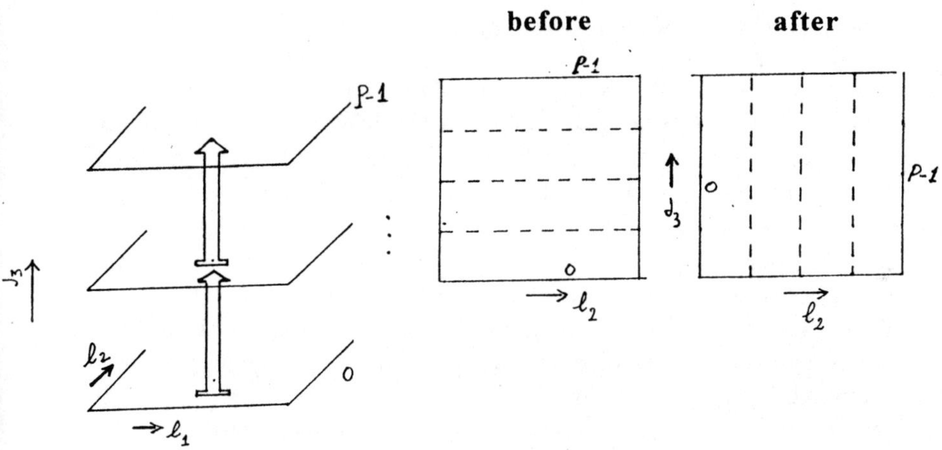

Figure 6.9: Data exchange (white arrows) flow to transpose data from (l_2, j_3) to (j_3, l_2) planes, denoted by before/after respectively

This is formally equivalent to a two-dimensional transpose

$$\hat{u}(l_1, l_2, j_3) \rightarrow \hat{u}(l_1, j_3, l_2) \ , j_3 = 0, 1, ..., N-1$$

i.e. an operation requiring all-to-all communication.

Once the transpose is available, again, each processor can complete N^2 one-dimensional FFT's along z in a completely independent way:

$$\hat{\hat{u}}(l_1, l_3, l_2) = FFT_z\{\hat{u}(l_1, j_3, l_2)\}; \ l_1 = 0, N-1 \ , l_2 = 0, N-1$$

As a final step, a back transposition is performed:

$$\hat{\hat{u}}(l_1, l_2, l_3) = Transpose\{\hat{\hat{u}}(l_1, l_3, l_2)\}$$

Summarizing, the parallel 3D transform consists of 4 steps:

1. N 2D-FFT transforms within Π_{xy}
2. Transpose $(k_y, z) \rightarrow (z, k_y)$
3. N^2 1D-FFT transforms along Λ_z
4. Transpose back $(k_z, k_y) \rightarrow (k_y, k_z)$

As a result, communication is concentrated in the transposition step. Several algorithms have been developed in the literature to achieve optimal efficiency. Here we shall sketch only the two simplest instances: the "dealing" complete-exchange algorithm and the "shuffle" exchange algorithm, which are suitable for bandwidth-bound and latency-bound interconnecting networks respectively.

Let us consider the case $P = 4$ for simplicity.

Within the full-exchange algorithm, each processor sends its parcel of data to every other processor over a series of $(P-1)$ communication steps. The sequence is sketched in Fig. 6.10.

This scheme completes the communication in $(P-1)$ steps, each of which involves the exchange of N^3/P (one plane) data from one processor to another.

In addition to requiring (P-1) steps, this scheme is not contention-free because some communication steps (first and third in our example) involve more than two processors simultaneously. This could cause undesirable traffic congestion effects which may tax the communication efficiency.

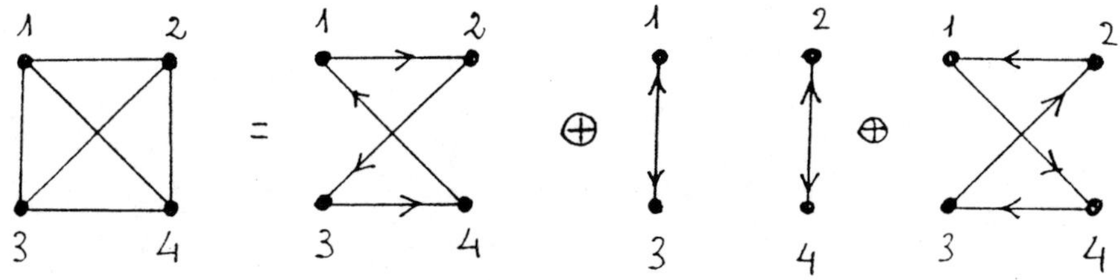

Figure 6.10: Pattern of the full-exchange algorithm to perform all-to-all communication among 4 processors

For most machines the "shuffle" schemes proves more advantageous. This is a sort of "clustered" communication scheme in which each processor serves as intermediate router to the final destination processor, thereby reducing the number of communication steps from $(P-1)$ to $\log_2 P$.

The scheme is sketched in FIG. 6.11.

The content of the messages is as follows:

$1 \rightarrow 2$; 12, 14
$2 \rightarrow 1$; 21, 23
$1 \rightarrow 3$; 13, 23
$3 \rightarrow 1$; 31, 41

$3 \rightarrow 1$; 34, 32
$4 \rightarrow 3$; 43, 41
$2 \rightarrow 4$; 24, 14
$4 \rightarrow 2$; 42, 32

where *pq* indicates the data that must be exchanged between processors *p* and *q*.

With this scheme communication is completed in $\log_2 P$ steps: in the first step only one plane of data is exchanged and in the second two. For the general case of P processors, the p-th step involves 2^{p-1} data planes, so that the total amount of planes exchanged is

$$\sum_{p=1}^{\log_2 P} 2^{p-1} = (P-1)$$

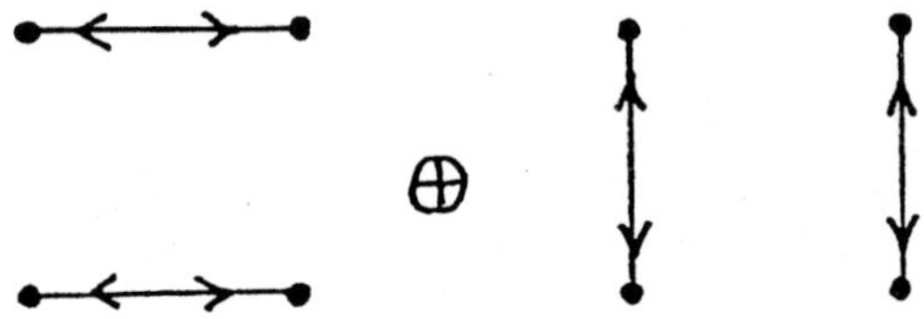

Figure 6.11: The shuffle-exchange pattern for all-to-all communication

as it should be.

This leads to longer messages and also to a contention-free scheme.

The algorithms previously outlined entail roughly

$$O(N^2 \ln_2 N^2) + O(N^3 \ln_2 N) \sim O(N^3 \ln_2 N)$$

operations per processor, and the exchange of

$$N^2(N-1)$$

items per processor (items are usually simple precision words, i.e. 4 bytes).

The net communication/computation ratio is of the order of

$$CCR \sim O(\frac{1}{\log_2 N})\text{words/flop}$$

which shows that SM are relatively "lean" compute-wise ($\log_2 N \sim 10$ for present-day largest supercomputers).

It is worth to point out a qualitative difference with respect to local grid methods. In the latter, communication versus computation always goes with "surface to volume", as it befits a local method.

The result is that the CCR no longer depends on the number of processors P, and consequently speed-up as a function P must proceed along a straight line until other types of overheads enter the scene.

The slope of this line is given by $(1+CCR)^{-1}$, and grows therefore very slowly with the problem size.

We remark that the present domain decomposition $X = \Pi_{xy} \otimes \Lambda_z$ implies that each processor must afford enough memory to host a full two-dimensional problem.

For leading-edge calculations ($N \sim 1024$), this entails $O(10MB)$ local-node storage, an amount currently supported by all most prominent parallel platforms.

Obviously, scalability is bound to $P \leq L$. This limit could be raised to P^2 by a two-dimensional partitioning $X = (\Lambda_x \otimes \Lambda_y) \otimes \Lambda_z$ without significantly affecting the CCR of the application.

It would however generate more data traffic (number of messages) and also potential latency exposures ($L = 1024$, at $S_1 = 40$Mflop/s, imposes a latency below $100\mu s$, since pencils -instead of planes- have to be exchanged between the various processors).

For these reasons most current parallel SM implementations are based on "plane and pencil" one-dimensional decompositions with $P = N$ [110].

6.5 Particle Methods

Particle Methods, as well as Lagrangian methods, exhibit a sort of natural potential for parallel computing which is explicitly built-in within their very formulation.

This formulation rests upon a sharp splitting of particle dynamics in two well-distinct steps:

- Free-propagation
- Collision

This sharp distinction, which is increasingly accurate with decreasing range of the inter-particle potential, permits to handle particle trajectories as basically independent entities and consequently to evolve them concurrently.

Moreover, particle dynamics is generally characterized by fairly low communication to computation ratioes, because most of the computational work is performed in the collision step. Being the collisions local in configuration space (short-range potential) little communication is required.

In some sense, particle methods can be paralleled to Explicit grid methods, with the "advantage" of **not** being computationally lean.

These are the two reasons why Particle Methods go hand in hand with parallel computing.

There are downsides as well. The main question is how much of this compute density is really needed to the purpose of the hydrodynamic description; much of the collisional work may well be indispensible only to the end of capturing physical effects laying beyond the fluid-level (see for instance the Direct Simulation Monte Carlo for rarefied gas dynamics).

So, the key issue is to define the minimal "hydrodynamic Lagrangian units", such that their dynamics be completely equivalent to the Navier-Stokes equations without carrying over any redundant information from the "upper-levels" of the BBGKY hierarchy.

To date, the Lattice Boltzmann equation seems to provide the closest instance of these "minimal Lagrangian units", to be identified with the discrete populations f_i of the lattice.

The hydrodynamic information associated with these (14 to 18) populations corresponds to 10 hydrodynamic fields only, namely density ρ, fluid speed $\vec{u}$ and six independent component of the stress tensor $\overset{\Rightarrow}{S}$.

Thus, a factor $1.5 \div 2$ redundancy is the price to pay to formulate the Navier-Stokes equations in terms of populations $\equiv$ "particles" moving along well prescribed trajectories.

It is quite possible that a "pathological" degree of redundancy must necessarily come along with particle formulations: whether or not leaving with this extra-formulation is a good bargain in view of a better amenability to parallel computation, still remains to be seen.

6.5.1 Parallel Molecular Dynamics

As discussed in Chap. IV, Molecular Dynamics (MD) is not yet to the point where it can be used as a competitive method to simulate large-scale complex flows. It can, however, shed important lightspots in the fascinating territory where fluid behaviour gives way to microphysics.

For this reason, it appears a natural candidate for multi(scale) physics codes merging atomistic and continuum description of complex physical phenomena often met in material science applications coupled with flow phenomena [77].

Parallelization of MD methods offers several interesting technical highlights, which we shall touch upon in the following.

The computational kernel of MD schemes is the computation of the force exerted upon molecule i by all other molecules $j = 1, N$:

$$F_i = -\sum_{j \neq i} \frac{\partial}{\partial x_i} V(|x_i - x_j|) \equiv \sum_{j \neq i} F_{ij} \tag{6.30}$$

Here, the potential $V(r)$ will be assumed short-ranged (Lennard-Jones-like). This implies that only a limited number of neighbors of molecule i do contribute to the force summation in eq. 6.30.

Parallel MD methods split into two broad categories [111] [112]

Spatial Decomposition (SD)

Particle Decomposition (PD)

depending on whether partitioning over the various processors applies to the configuration space $\vec{x}$, or rather the discrete degrees of freedom carried by the molecules in phase space $\Gamma \equiv \{\vec{x}_i, \vec{v}_i \, ; i = 1, ..., N_P\}$.

In a recent time, a third possibility has emerged, which consists of direct partitioning of the force matrix F_{ij}. This will be referred to as to Force Decomposition (FD) [113].

Spatial Decomposition

Let us begin with SD. In this method, each processor p is assigned a spatial subdomain Ω_p of the configuration space and is therefore responsible for the dynamic update of all particles whose coordinates fall within Ω_p (see Fig. 6.12 for a simple one-dimensional decomposition in $D = 2$ - Eulerian decomposition).

This method bears close resemblance to grid-partitioning techniques, and more specifically, with **explicit unstructured** ones. As a result, they share advantages and concerns, i.e., load-unbalance exposures, both in computation and communication, laborious bookkeeping, latency exposures due to short messages. In return, they provide potentially good scalability if all the above issues can be kept at bay.

Let us comment in some more detail.

Load unbalance exposures stem from the fact that an even distribution of the size of the subdomains Ω_p does **not** imply a correspondingly even

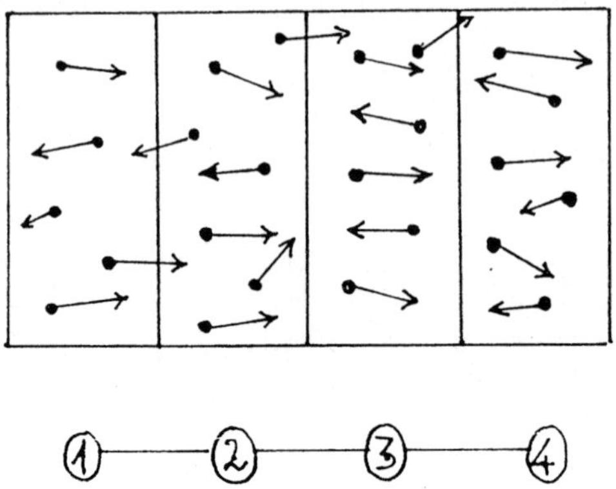

Figure 6.12: One-dimensional spatial decomposition of a two-dimensional domain. Particles flying across interprocessor boundaries trigger data exchange

distribution of the number of particles N_p belonging to Ω_p. This is especially true in highly non-homogeneous systems with sudden density changes, such as compressible flows.

This calls for dynamic load balancing strategies responding to the changes of N_p with a corresponding change of the size of Ω_p.

A similar issue applies to communication. Interprocessor communication is triggered whenever a particle flies across an interprocessor boundary.

Again, a uniform distribution of the size of interprocessor boundaries does **not** ensure a corresponding uniformity of the flux of particles across them.

A further issue is geometrical complexity, as amply discussed for grid methods, finding a "good" decomposition (independently of the dynamic concerns previously mentioned) when the geometry is complex can be a difficult task even under **static** conditions.

This is a price to pay for having reintroduced a grid, even though only for parallelization purposes, into an otherwise gridless dynamics.

The return, as anticipated, is potential good scalability. Let us see why: the argument goes exactly like for explicit structured methods.

Under uniform conditions, calculations (to advance from t to $t + \Delta t$)

within a single processor scale like

$$T_{cal} \sim \chi N_P/P$$

where χ is the computational density (flop/particle), N_P the total number of particles and P the number of processors.

Within a d-dimensional partitioning, communication scales like

$$T_{com} \sim \sigma(N_P/P)^{1-\frac{1}{d}}$$

where σ is the communication density (number of particles flying across interprocessor boundary). As a rough estimate:

$$\sigma \sim \bar{v}t, \quad \chi \sim b$$

where $\bar{v}$ is the average molecules speed and b the average number of neighbors of a given molecule.

As a result, we obtain

$$CCR = \frac{\bar{v}}{b}(\frac{P}{N_P})^{\frac{1}{d}}$$

which ensures good scalability for $P < P_{crit} \sim N(b/\bar{v})^d$. In practice, this asymptotic regime may be very hard to attain because of the non-idealities previously discussed.

Whenever these non-idealities are too restrictive, Particle Decomposition (PD) may come to (partial) rescue. Within PD, each processor is assigned a fixed subset of particles Γ_p and keeps tracking it all along the simulation (Lagrangian Decomposition).

The advantages are clear: first, load-balancing is secured right at the outset, since the number of particles assigned to each processor is constant in time. Second, grid-freedom is automatically carried over from the serial scheme since the partitioning is totally insensitive to the geometry of the problem. Third, the book-keeping burden is minimal.

The price to be paid for this simplicity is that global communication may arise as a result of particle wandering in space, which may jeopardize the potential for full scalability.

Let us discuss this point in a more detail.

Denoting by Ω_p the spatial domain associated with the subset Γ_p of molecules owned by processor p, it is clear that, under the effect of particle

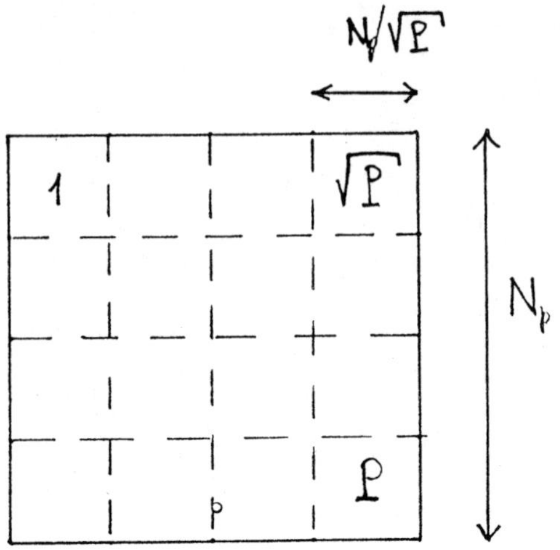

Figure 6.13: Lagrangian Particle Decomposition. Under the effect of particle dynamics initially separate subdomains merge and mix thus generating highly complex communication patterns

dynamics, Ω_p can undergo significant distorsions and changes of shape (see Fig. 6.13).

The net result of these distorsion is that originally well separated subdomains Ω_p, Ω_q may well intersect in the course of time, thus triggering interprocessor communication among processors p and q.

With the same token, multiple intersections $(\Omega_p \cup \Omega_q \cup \Omega_r .. \neq 0)$ can occurr which ultimately may end up into a sort of random global communication scheme:

It is this global communication that kills scalability. In fact, under the conditions of Fig. 6.13 the communication requirements no longer scale like a "surface" term $(N_P/P)^{1-1/d}$, but rather like a "volume" term (N_P/P), so that the resulting CCR is significantly higher than for SD in the scaling region $P < N_P$

Despite these shortcomings, PD methods can offer a valid alternative to SD whenever the number of processors is not too high and the geometry to be modelled is complex enough.

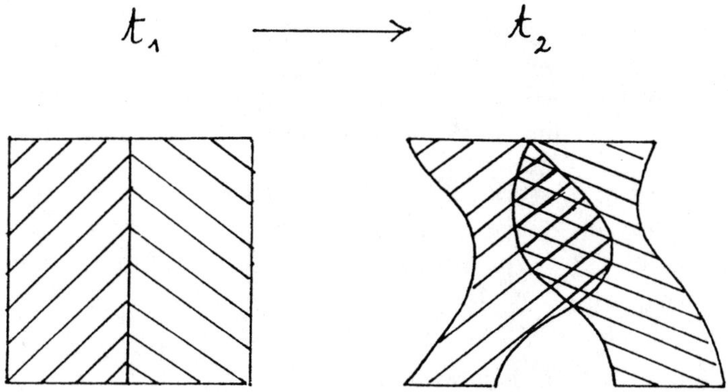

Figure 6.14: Two-dimensional partitioning of the force matrix induced by the splitting of the molecular positions

Force Decomposition

Force decomposition (FD) aims at combining the merits of SD and PD while leaving their drawbacks apart.

Let us see how. First, assume the array of molecular positions $x = \{\vec{x}_i,\ i = 1, N\}$ is split into $\sqrt{P}$ groups of length $N/\sqrt{P}$; $x = x^{(1)} \oplus x^{(2)} \ldots.. \oplus x^{(\sqrt{P})}$.

This induces a two-dimensional partition of the force matrix into P sub-blocks, as shown in Fig. 6.14 ($P = 16$).

The $\sqrt{P}$ processors within a row further subdivide their group in such a way that each processor takes care of only N_P/P molecules, while owning at the same time the whole set of coordinates of the molecules assigned to its row-mate processors (i.e. $N_P/\sqrt{P}$ coordinates) and column-mate processors ($N_P/\sqrt{P}$ coordinates).

The process of computing the global force F_i acting upon the molecule i, is equivalent to a (random) sparse matrix-vector-multiply (mvm) of the form

$$F_i = \sum_{j \neq i} F_{ij} 1_j \tag{6.31}$$

where F_{ij} is generally sparse and skew-symmetric ($F_{ij} = -F_{ji}$) due to Newton's third law and 1_j is a constant array of ones.

Parallelization of the sparse "mvm" proceeds in six steps

1. Each processor computes F_{pq} in its block ($p, q = 1, \sqrt{P}$). For the example of Fig. 6.14, proc. 1 computes F_{11}, proc.2 F_{12}, i.e. the force exerted by molecules in $x^{(2)}$ upon molecules in $x^{(1)}$ and so on.

2. Each row of processors **folds** the force vector, i.e. every processor in the row forwards its force contribution to all of its row-mates. At the end of the "fold", each processor knows the total force acting upon its share of particles:

$$F^{(p)} = \sum_{p \neq q} F^{(pq)} + F^{(pp)}$$

3. Each processor updates the position of its N_P/P molecules.

4. Each row **expands** the position array: i.e. every processor in the row receives the positions of the molecules updated by its row-mates processors (the one it owns so far, i.e. up to step 3, would not be up to date). As a result, each processor knows the updated coordinate for its entire row group.

5. Each processor exchanges N_P/P particle positions with its transpose in the force matrix and stores them in its column position array.

6. Each processor "expands" its position array across all of its column mates. As a result, each processor knows the updated position for its entire column group.

7. A new step can then begin.

Although rather laborious, the FD method is appealing because: first, it preserves grid-freedom: no information on the underlying geometry is needed at any point.

Second, load-balancing issues are softened because if molecules are ordered randomly, the sparsity pattern will be (relatively) uniform over time, so that each block $F^{(pq)}$ should approximately contain the same number of non-zero entries.

Third, it scales better than PD because the communication "fold" and "expand" primitives can be performed in parallel (over $\sqrt{P}$ groups of processors) in a time proportional to the length of the vector being folded (expanded), i.e. $N_P/\sqrt{P}$, even if the amount of data exchanged is still "volume-like", $O(N_P)$, like in PD.

A comparative study of SD, PD and FD methods on distributed multiprocessors (Intel and Ncube) [113] shows that FD can compete in performance with SD for up to tens of thousands particles on $O(10^3)$ processors. However, for multimillion particle simulations, such is the current state-of-art, SD outperforms FD by a factor $2 \div 3$.

6.5.2 Parallel Direct Simulation Monte Carlo

The parallelization of Direct Simulation Monte Carlo (DSMC) algorithms proceeds pretty much along the lines described for Molecular Dynamics (MD) schemes. As in MD, there are two basic options: spatial decomposition (SD) and particle decomposition (PD).

SD ensures better scalability in that it preserves locality in the communication pattern, but it presents serious exposures to load unbalance if the particle populations happen to distribute unevenly across the spatial cells. Of course, load unbalance can be leveled-off by dynamically readjusting the spatial grid to the actual computing load, but this is a very expensive operation.

With PD load-balance is automatically secured, but only at the expense of a potential inflation of the communication pattern (particles assigned to different processors meeting within the same spatial cell).

This loss of locality reflects in the need of sorting particles at each time step according to the index of spatial cell they lie within. This sorting step may well be the bottleneck of parallel DSMC codes even on machines catering for special-purpose micro-coded routines designed to rank disordered sets [114], [115].

Being DSMC targeted to rarefied gas dynamics, a discipline in which the main trigger of load unbalance, i.e. sharp density excursions, is a commonplace, one may argue that PD is the right option.

This is however by no means a clear-cut argument, and the ultimate choice may well depend on architectural details, as well as on the complexity of the physical phenomena to be included in the simulation [114].

6.5.3 Lattice Gas

Lattice Gas Cellular Automata (LGCA) can be classified as **boolean** explicit structured schemes.

They offer three distinct levels of parallel potential [116]:

1. Bit level
2. Discrete speed level
3. Grid level

The bit-level parallelism refers to the opportunity -unique to boolean schemes- of updating 32 (64) bits at a time by means of a single computer instruction.

This technique -known as **multispin coding** [117]- was is use since more than a decade within the Statistical Mechanics community, but it was only with the advent of LGCA that it entered the realm of CFD.

As we have seen, LGCA dynamics proceeds via two basic evolution steps: free-streaming and collision.

Multispin coding of free-streaming is very straightforward. Suppose 32 rightward propagating particles are packed into a single computer word, say, R; the "Move-right" free-streaming reads simply:

$$R(t+1) = r\_\text{shift}(R(t), 1)$$

where $r_$shift is a pseudo-language notation for the instruction shifting the bits of a computer word one unit rightwards. (In actual practice, some care is needed to handle the boundary bits correctly).

Similarly, the collision step can be implemented as

$$\begin{aligned} R' &= Colli_R(R, UR, UL, L, DL, DR) \\ &\vdots \\ DR' &= Colli_{DR}(R, UR, UL, L, DL, DR) \end{aligned} \tag{6.32}$$

Here $R, ..., DR$ stand for the six discrete speeds defined in the FHP lattice "right, up-right, up-left, left, down-left, down-right" respectively.

Colli_xx is a pseudo notation for the collision operator which can be constructed as an appropriate combination of elementary boolean operators.

Discrete-speed parallelism may be viewed as a form of functional parallelism applicable to the free-streaming phase. As a such, it is peculiar to discrete-kinetic theory. The idea is simply to let each processor moving bits along a distinct direction in a fairly independent way.

In other words, processor 1 may move bits "right" at the same time as processor 2 moves "up-right", processor 3 "up-left" and so on.

This is trivially programmed on shared-memory MIMD computers, but doesn't extended easily to distributed memory machines.

In addition, its scalability is upper-bounded by the number of discrete speeds, i.e. 6 in two-dimensions and 24 in three.

The parallelism available at the grid level is the one we already accustomed to after having discussed explicit finite-differences.

Yet, the peculiar nature of LGCA reflects in an especially high compute density (the Collision phase) as compared to the communication density (the Move phase, which is first-order in space). More precisely, the compute density scales exponentially with the number of discrete speeds, b:

$$\chi \sim 2^b$$

while the communication density scales only linearly

$$\sigma \sim b$$

The result is an outstanding low CCR, which characterizes LGCA as almost "communication-free" explicit schemes:

$$\frac{\sigma}{\chi} \sim b2^{-b} \sim 0; \ (b > 6)$$

This is in a blatant constrast with "classical" explicit CFD schemes for which,as repeatedly pointed out, σ/χ is $O(1)$.

In that respect, it is hard to imagine anything more suited to parallel computing than LGCA. The problem is, however, that optimal parallel efficiency is a meager consolation when faced to an exponential escalation of the compute density.

This exponential escalation -together with statistical noise- is still the main hurdle preventing a significant penetration of LGCA within the CFD community.

6.5.4 Lattice Boltzmann

The discussion of parallel Lattice Boltzmann schemes follows the main issues outlined for LGCA, with two important distinctions:

1. Bit-level parallelism is no longer available

2. Compute density is $O(b^2)$ instead of $O(2^b)$

As to 1), the loss of bit parallelism is no big issue since it is more than compensated by the removal of statistical noise. Regarding point 2), LB is somehow optimal in the sense that it brings down the compute density to a relatively acceptable level while keeping a $CCR \sim O(1/b)$ still significantly lower than most "conventional" explicit CFD schemes.

As a result the Lattice Boltzmann method is an extremely good parallel performer on virtually all computer platforms.

6.5.5 Vortex Methods (VM)

Vortex methods stand out as appealing discretizations for high-Reynolds unsteady flows, their primary merits being grid-freedom and automatic self-adaptivity, as discussed in Chapter IV.

Owing to the long-ranged nature of the inter-vortex potential, VM fall within the important computational class of **numerical N-body problems**, which find many-faceted applications in physics and engineering [118] [119]

The dominant kernel of the numerical N-body problem is the computation of all pair-wise interactions which concurr to form the force F_i acting upon the i-th body (vortex):

$$F_i = -\sum_{j \neq i} \frac{\partial}{\partial x_j} V(|x_i - x_j|) \equiv \sum_{j \neq i} F_{ij} \tag{6.33}$$

Numerical N-body methods come in two broad classes [120]:

Direct Summation Techniques (DST)

Fast Summation Techniques (FST)

a distinction that applies to their parallel counterpart as well.

Direct Summation Techniques lend themselves to straightforward data parallelization over the set of vortex degrees of freedom.

Grouping $n = N/P$ vortons within the same processor, the global interaction dynamics leads to the following requirements:

Intra-processor calculation: $O(n^2 P)$

Inter-processor communication: $O(nP)$

with a resulting computation/communication ratio

$$CCR = O(n^{-1}) \tag{6.34}$$

The eq. 6.29, simple as it is, spawns a number of remarks: First, it shows that high parallel efficiency are readily achievable thanks to the extreme compute intensity of the method. Taking the threshold of acceptable efficiency at $n = 1$, we obtain $P \ll N$. Since N is typically $O(10^4 \div 10^5)$ in current simulations, the method can definitely be regarded as a scalable one.

Second, note that eq. 6.29 holds independently of the dimensionality of physical space, as well as any surface/volume-like parameter.

This is a direct consequence of the global nature of the interaction combined with the grid-freedom of the technique which rules out any geometric domain-decomposition parameter.

Owing to these simple, and yet powerful, features, direct summation is a popular technique for parallel implementations; so popular as to justify the construction of special-purpose machines.

In spite of this, one should concede that living with $O(n^2)$ complexity is probably not the best choice, given the fact that serial algorithms exist which take the quadratic complexity down to $O(n \log_2 n)$ or even $O(n)$ operations.

These algorithms are too sophisticated to be described here; all the same, the general ideas are worth being rapidly summarized.

The key insight leading to FST predates parallel computers by some three-hundred years, for it can be traced back to Isaac Newton in 1687. The idea goes as follows: since the intensity of the interaction between particles falls off rapidly with distance, the effect of a sufficiently large group of particles can effectively be replaced by a single, representative, macroparticle, provided that this macroparticle is far enough from the point where its effects are being evaluated.

The practical hint stemming from this observation is to replace (certain portions of) the direct summations in eq. 6.33 by appropriate series expansions, usually known as **multipole expansions** [119].

With reference to the situation depicted in Fig. 6.15, a typical multipole expansion of the vortex potential in $D = 2$ is given by:

$$\phi_i(\vec{x_1}, ..., \vec{x_N}) = \sum_{j \neq i} |\vec{\Gamma_j}| \log(|\vec{x} - \vec{x_j}|) \sim$$

$$\Gamma^{(0)} \log |\vec{x}_i| - \frac{\Gamma^{(1)}}{|x_i|} + O(\frac{R^2}{R_i^2}) \tag{6.35}$$

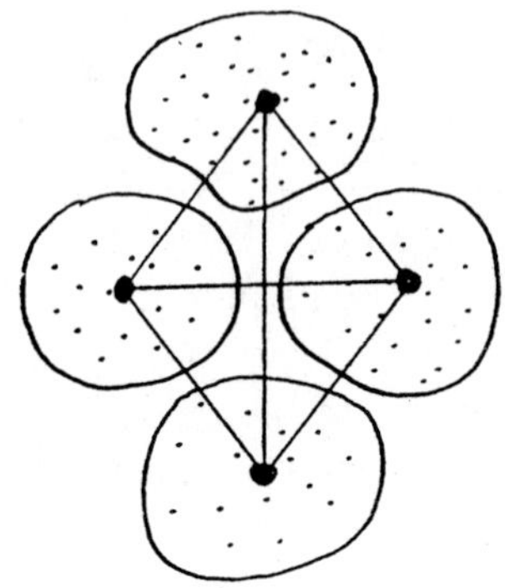

Figure 6.15: The interparticle interaction among a group of 4 macroparticles

where $\Gamma^{(0)} = \sum_i |\vec{\Gamma}_i|$ is the macro-vortex intensity and $\Gamma^{(1)} \equiv \sum_i \vec{\Gamma}_i \cdot \vec{x}_i$ is the dipole moment; R is the macroparticle separation and R_i the macroparticle radius (Fig. 6.16).

The expression 6.35 is a typical **far-field** approximation to the exact N-body potential, which can be profitably used to compute the interaction between particles lying sufficiently far apart.

The computational savings are obvious: once $\Gamma^{(0)}$ and $\Gamma^{(1)}$ are computed for each macroparticle $\mathcal{M}_p$, $p = 1, P$, the effects of $\mathcal{M}_p$ over the particles belonging to $\mathcal{M}_q$, $q \neq p$ can be computed via the expression 6.35 within $O(n)$ operations.

This brings a qualitative change in the communication-computation budget:

Computation	1. $n(n-1)/2$	intra-processor direct interaction
	2. $2n$	form $\Gamma^{(0)}, \Gamma^{(1)}$
	3. $(P-1)n$	inter-processor far-field interaction
Communication	$2(P-1)$	items $(\Gamma_q^{(0)}, \Gamma_q^{(1)}\ ; q \neq p)$

The corresponding CCR is given by:

$$CCR \sim O(\frac{1}{N^2/P + N + N/P}) \tag{6.36}$$

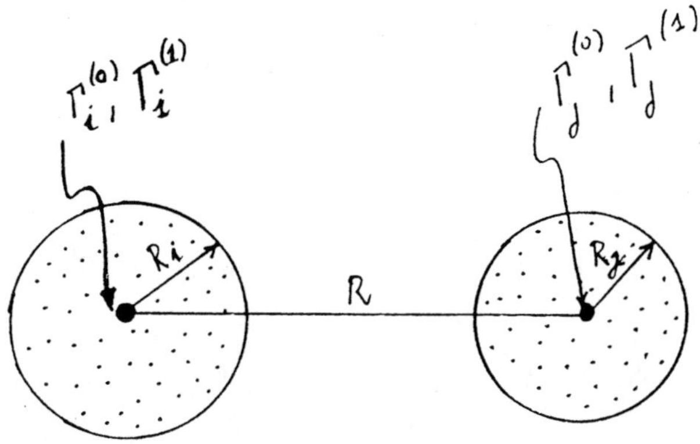

Figure 6.16: The geometrical parameters governing the multipole expansion

and comes eventually to the favourable $O(P/N)$ scaling for $P > N^{1/2}$, while being dominated by a $O(N^2/P^2)$ compute complexity rather than $O(N^2/P)$ like FST.

This transparent picture becomes a bit blurred in practical simulations where the distinction between sample and source points may be not so sharp.

A useful book-keeping device to keep this distinction well-defined across the calculation is to organize the interaction into a hierarchical structure, leading to a **tree-code** approach [119].

In its simplest instance, a serial tree-code introduces a hierarchy of boxes which refine the computational domain into smaller and smaller regions corresponding to increasing levels of resolution $l = 0, 1..., \mathcal{L}$ (see Fig. 6.17).

The basic idea is to define clusters of particles at successive levels of spatial refinement, and compute the interaction between "distant" clusters by means of far-field approximation, while neighboring clusters are computed by direct summations.

Clearly, the ratio of "distant" to "neighboring" clusters grows with the refinement level, so that for l large enough ($l \gg 1$) we can assume that, at each level, the compute complexity is $O(N)$.

On the other hand, since in D dimensions there are $\mathcal{L} = (1/D)\log_2 N$ levels, the global complexity becomes $O(N \log_2 N)$ as expected.

The operation count can be further cut down to $O(N)$ by subtle insights

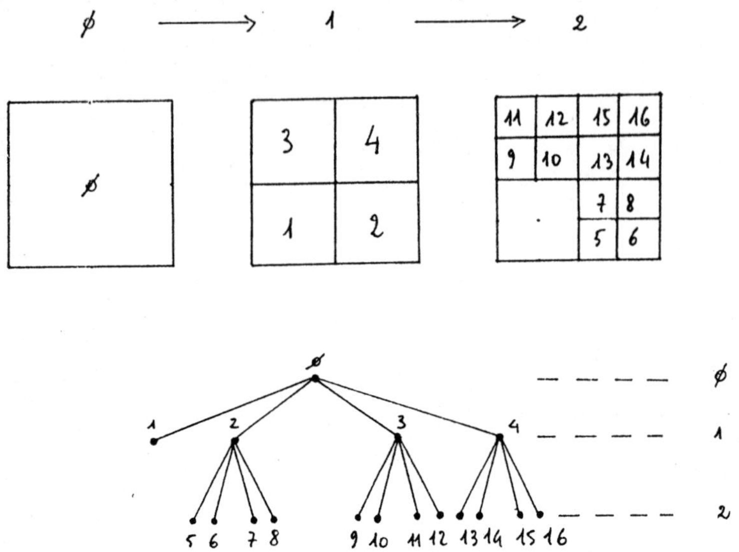

Figure 6.17: The hierarchical data structure associated with 3 levels of successive refinement of the computational grid

into the physics of N-body problems, which permit to compute the expansion coefficients at level l directly from those of $(l-1)$-th level (merging expansion) and viceversa (shifting expansion).

Here "expansion" is, in fact, a double expansion in which the original far-field expansion pertaining to the "source" macroparticle $\mathcal{M}_p$ is further re-expanded around the centroid of the "target" macroparticles $\mathcal{M}_q$ ("duality principle").

These ideas lead to the so called **F**ast **M**ultipole **M**ethods (FMM). At the first level, the FMM work is $O(N)$. The cumulative amount of work at all coarser levels, however, no longer depends on N, but rather on the number of macroparticles P.

Since in a tree of depth $\mathcal{L} = \log_2 N$ there are $O(N)$ leaves, the cumulative work of FMM is $O(N)$.

Keeping a good parallel efficiency for tree-codes and FMM codes is one of the forefronts of modern computational physics. Without entering the details of this rapidly evolving subject, let us list the main barriers.

Tree codes reintroduce spatial-decomposition concepts, hence vortex-to-processor mapping issues somehow similar to those encountered for unstructured grid methods. Here, potential exposures to inefficiency and possibly even failures are even higher. The reason being:

1. Load unbalance: Equal number of vortices does not imply equal work, since larger vortices require more updates. This can be paralled to difficulties encountered in numerical combustion where different regions may be exhibit wildly disparate chemical activity in the course of time. Since the cost of the cell is not known *a priori*, **dynamic load balancing** is called for.

2. Defective vortex-processor assignement not only will result into higher CCR's, but may even inficiate the validity of the multipole expansion itself, thus leading to code failure.

Despite these difficulties, successfull implementations on various parallel architectures are being reported in the literature for both MIMD (the most natural choice) and SIMD programming models [120]

Chapter 7

Parallel Implicit Methods

As discussed in Chap IV, implicit schemes are the method of choice whenever the main issue is to march in large steps towards the desired solutions with no need to describe the dynamics of the smallest scales of flow motion. Implicit methods require linear algebra solvers; these split into two broad categories: **direct** and **iterative**.

Direct solvers face the global dependencies of the inverse matrix A^{-1} head-on, while iterative solvers break this global dependence into a series of local iteration steps. For this reason, they are generally regarded as more suitable to parallel computing [121].

In the recent years, however, significant progresses have been achieved also in the area of parallel direct solvers.

Let us begin by discussing parallel iterative schemes.

7.1 Parallel iterative schemes

We shall consider two families of iterative schemes [122] :

- **relaxation** methods
- **gradient** methods.

7.1.1 Relaxation schemes

With regard to the linear algebraic problem $Af = g$, we shall consider iterative schemes in the following general (relaxation) form:

$$(I - A_1)f^{k+1} = (I + A_2)f^{(k)} + g^{(k)} \tag{7.1}$$

where A_1, A_2 represent an appropriate splitting of the matrix A.
Let us start with the simple Jacobi iteration scheme:

$$A_1 = D \ , A_2 = L + R$$

or, more explicitly:

$$f^{k+1} = (I - D)^{-1}\{g^k + (L + R + I)f^k\} \equiv \\ Jf^k + (I - D)^{-1}g^k \tag{7.2}$$

where J is the Jacobi "iteration-time" transfer operator.

We immediately recognize that eq. 7.2 has exactly the same form of an explicit scheme marching along the "fake" iteration time direction labelled by the index k.

As a such, the basic kernel to be parallelized is again the sparse matrix-vector scalar product discussed in the previous chapter.

Although a prototype of "easy" parallelization, Jacobi schemes are of limited practical use mainly because of their slow convergence.

They nonetheless serve as a sort of "elementary brick" for other, more practical relaxation schemes.

To this regard, let us consider the Gauss-Seidel (GS) relaxation scheme, characterized by the choice: $A_1 = L + D$, $A_2 = R$:

$$[I - (L + D)]f^{k+1} = (I + R)f^k + g^k \tag{7.3}$$

Here, the new feature is that, even if no linear algebra solver is required, the matrix L introduces a recursive dependence which prevents an explicit parallelization "à-la" Jacobi.

In the context of parallel computing, recursion is most conveniently viewed as a **delay-dependence** (the matrix L is *de facto* a rightward propagator) whose effect is to "freeze" the status of unknown f_n until all predecessors $f_1, ..., f_{n-1}$ have been updated.

Let us clarify with a simple example: a one-dimensional chain with nearest-neighbor connections (tridiagonal matrix) (see Fig. 7.1).

The logical updating sequence is (assuming periodic boundary conditions):

$$4, 2 \rightarrow \hat{1} \ ; \ \hat{1}, 3 \rightarrow \hat{2} \ ; \ \hat{2}, 4 \rightarrow \hat{3} \ ; \ \hat{3}, 1 \rightarrow \hat{4}$$

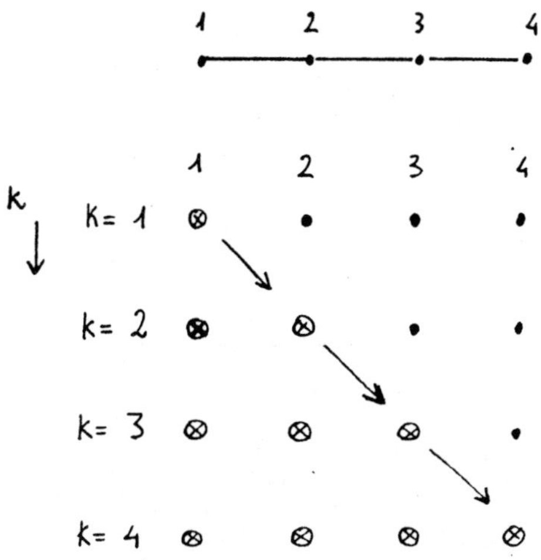

Figure 7.1: The updating sequence associated with GS relaxation. The i-th unknown can only be updated after i iteration sweeps

where caret denotes updated values at iteration level $(k+1)$. Thus, the rightmost site, 4, suffers a delay of 4 updating cycles.

An effective remedy to delay-dependence is **multicoloring** (MC) [123]. The idea of MC is to split the unknown array f into a series of disjoint sets ("colors") such that each color only depends on a different one. Once this condition id fulfilled, GS can proceed as a series of NC Jacobi "monochromatic" steps of the form:

$$(I + D_c) f_c^{k+1} = f_c^k - \sum_{c'<c} L_{c'} f_{c'}^{k+1} - R_c f_c^k + g_c^k \equiv g_c^{k+1}$$

$$c = 1, NC \qquad (7.4)$$

The idea is that whenever a color has been updated, the corresponding L matrix (L_c) is taken to the r.h.s. precisely like the R matrix, so that each color can be updated in parallel. Multicoloring is clearly a further approximation, which may lead to a degradation of convergence with respect to the serial, monochromatic GS. As usual, the assumption is that this slowing-down is eventually offset by the capability to proceed in parallel.

With reference to the one-dimensional case with nearest-neighbor interaction, it is readily seen that just two colors (say, red and black) will do.

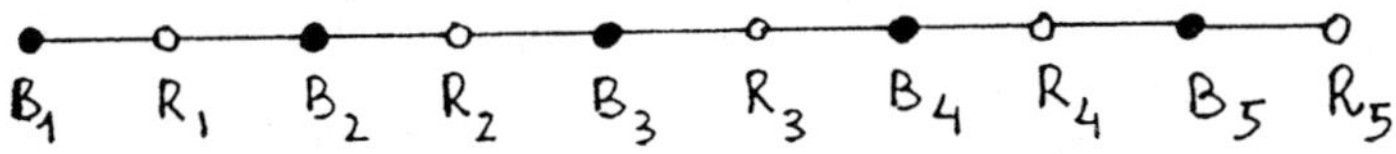

Figure 7.2: Red-black ordering of a chain. With nearest-neighborin connection, the red-subset only depends on the black and viceversa

The updating sequence and the corresponding numbering in Fig. 7.2 should be self-explanatory.

Two color schemes are mostly used for two-dimensional five-point stencil applications (see Fig. 7.3)

Note that red/black ordering is equivalent to numbering along left (right) handed diagonals respectively; in a way, it's a sort of "staggered" domain decomposition.

Another popular colored GS scheme is the 4-color (white, orange, red, black) scheme for two-dimensional, nine-point stencil application (Fig. 7.4).

This is equivalent to two pairs of right (left) handed "zig-zag" zebra patterns.

The problem of finding the minimum number of colors required to dissect a grid with a general stencil is equivalent to the graph coloring problem, hence NP-complete.

Multicolored schemes are potentially exposed to **synchronization overheads.** Consider in fact the updating sequence (inner iteration) consisting of four monocromatic Jacobi sub-iterations. In symbolic form:

$ORB \rightarrow \hat{W}$

$\hat{W}RB \rightarrow \hat{O}$

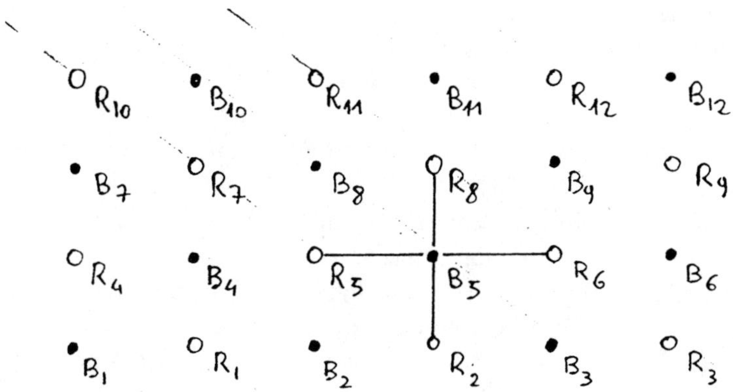

Figure 7.3: Red-black ordering of a two-dimensional regular grid

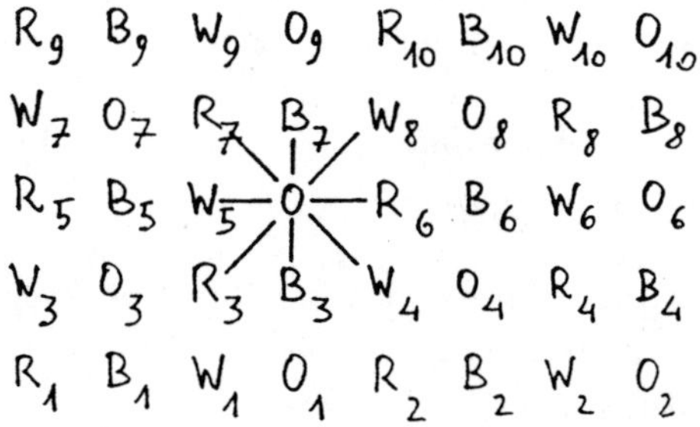

Figure 7.4: Four-color (red - black - white - orange) scheme for two-dimensional, 9-point stencil, applications

$$\hat{W}\hat{O}B \rightarrow \hat{R}$$

$$\hat{W}\hat{O}\hat{R} \rightarrow \hat{B}$$

It is clear that in order to be consistent, all processes should synchronize before each of the four Jacobi sweeps.

Should this result in an unacceptable penalty (high latency machines) one can resort to *stochastic relaxation* [124].

This is an elegant substitute for just "don't care": in other words, no synchronization is performed and each processor just proceeds with the most recent available values even if they're not up-to-date.

This obviously introduces "time misalignment" errors which cause deviations from the correct, synchronous scheme.

The point is that, hopefully, this misalignment does not change the final state, nor does it slow down convergence to this state.

Synchronization overheads show up at completion of each inner iteration, when the residuals accumulated by every single processor must be summed up to produce a global residual needed by the convergence check.

If latency is not an issue, a simple "all-to-one" ("gather" or "collect") technique can be used. Each processor forwards its residual r_p to, say, processor 0, which produces the global residual $r = \sum_{p=0}^{P-1} r_p$.

Typically this should result in a synchronization overhead of the order of $(P-1)\Lambda$, where Λ is the processor latency.

If this delay is unacceptable, one can implement a binary-tree communication scheme which should complete in $\ln P$ latency units (Fig. 7.6).

Synchronization overheads also show up at the completion of each outer iteration, the one dealing with the non-linearity.

Generally speaking, each time step involves basically three loops (see Fig. 7.7)

The total number of synchronization points per time-step is given by

$$N_s = (NC \cdot NIN) \cdot NOUT$$

where $NOUT$ is the number of outer iteration/step, NIN the number of (multicolor) inner iterations per outer iteration, and NC the number of colors of the inner iterator.

If, as is often the case, most of the work is performed in the linear solver, then one can focus efficiency considerations on the inner iteration loop, so that the proper count is just NC synchronization points.

More generally, especially for finite-volume and finite-element applications, each time step involves two basic units of computational work:

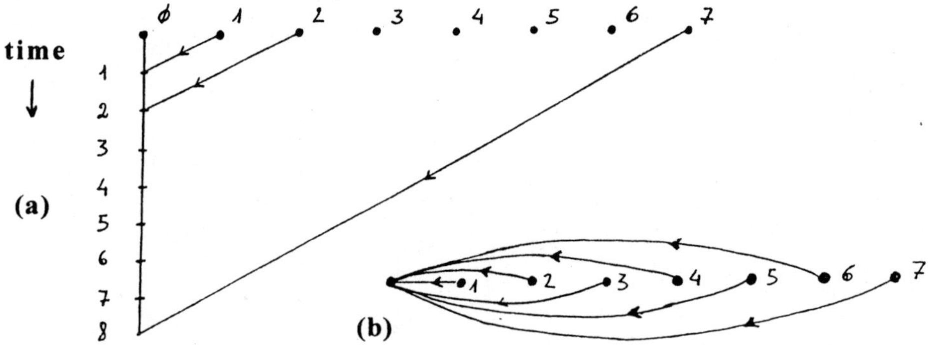

Figure 7.5: Time-line (a) and communication pattern (b) of the "all-to-one" implementation of the global communication required at completion of the iterative step

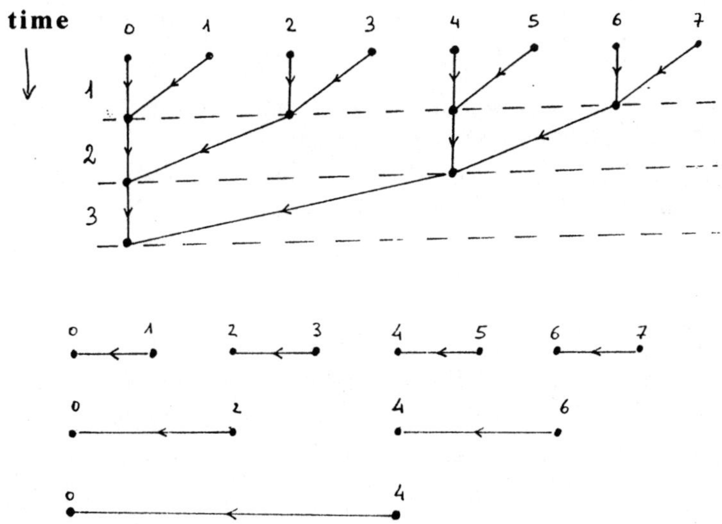

Figure 7.6: Time-line (a) and communication pattern (b) of the binary-tree implementation of the global communication

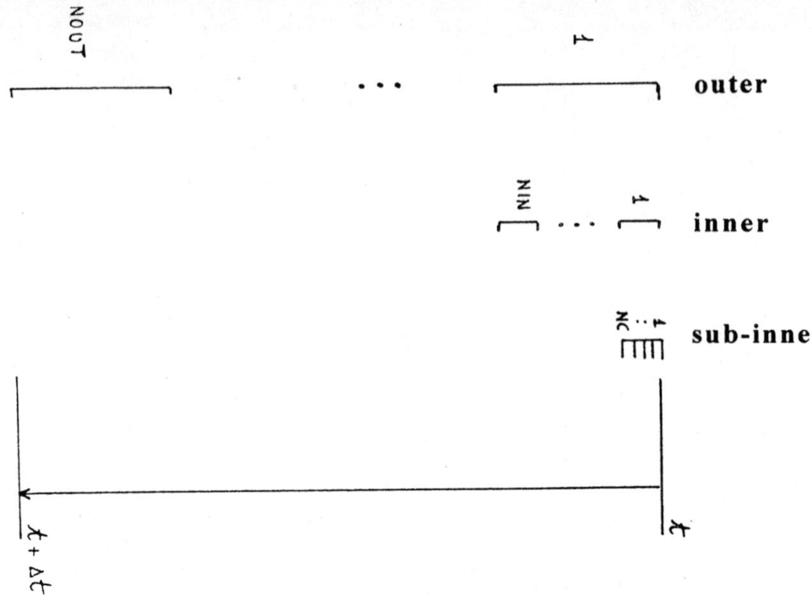

Figure 7.7: The three iteration loops: outer, inner and sub-inner required to advance the solution one step in time

a Formation of the matrix A and r.h.s. b

b Solution of $Ax = b$

Note that A changes from one outer iteration to the other because of the nonlinearity of the Navier-Stokes equations.

Calling G_A and G_S the "grains" of the assembly and solver steps respectively, the number of operations per time-step is given by

$$G = (NIN \cdot G_s + G_A) \cdot NOUT$$

This shows that if the number of internal iterations is sufficiently large, the overall CPU demand of the application is dominated by the solver.

As a rule of thumb $NIN \sim O(L)$, L being the linear size of the problem, so that, for large CFD problems, ($L \sim 10^2$) indeed solver dominance should be the rule.

7.1.2 Multicolored Block-Iterative Methods

The multicolored methods discussed so far work at the level of the single grid-point and are therefore referred as to **point** relaxation methods. In

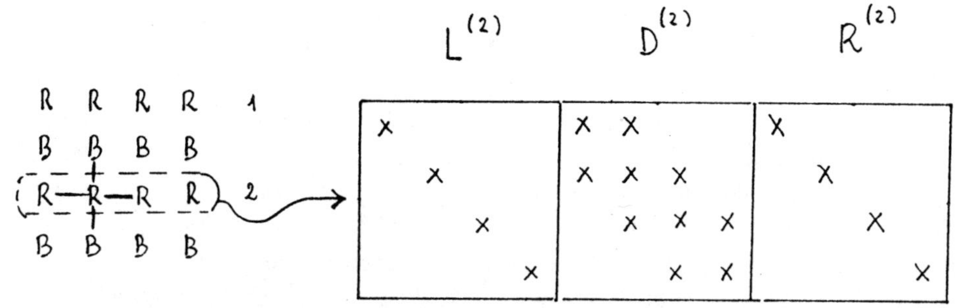

Figure 7.8: Red-black row-ordering of a two-dimensional domain. All variables in the same row are solved simultaneously with a direct method. Different rows are coupled iteratively.

order to speed-up convergence, at least on serial computers, point relaxation schemes are very often replaced by more tightly coupled "block" relaxation schemes in which entire subdomains are given the same color. The is idea that variables in the same block are tightly coupled and they should therefore be solved with "exact" direct methods. On the contrary, off-block coupling is assumed to be weak enough so as to justify the use of an "approximate", iterative, technique.

A simple example of multicolored block-iterative method is the line-relaxation two-color scheme depicted in Fig. 7.8

Assuming a 5-point nearest-neighbor stencil, this scheme leads to a series of block tridiagonal (BT) problems of the form:

$$
\begin{aligned}
D_j x_j^{k+1} &= -(L_j x_{j-1}^k + U_j x_{j+1}^k) + b_j \\
&j = 1, NY;\; k = 1, 2, ..., NIN
\end{aligned} \tag{7.5}
$$

where D_j is a block tridiagonal matrix of size (NX, NX) expressing the coupling between elements in the j-th row (x_j) and L_j and U_j are the (diagonal) matrices expressing the coupling of the row vector x_j with its lower and upper neighbors, x_{j-1}, x_{j+1}.

This scheme achieves faster convergence because only the extra-diagonal links are treated iteratively, while the couplings between nodes in the same row are solved "exactly" by a direct method.

In this respect, it is apparent that block iterative methods are in fact a blend of iterative and direct methods.

The parallel scheme proceeds into two sweeps, each of which consists of the simultaneous solution of $NY/2$ block tridiagonal problems, once for the odd and once for the even rows of unknowns.

These schemes are attractive mainly due to their conceptual and operational simplicity. However, they offer limited scalability and rather unconvenient surface/volume ratios (each node has to send its piece of solution to form the r.h.s. of its two neighboring nodes). This is why these schemes are most conveniently used for shared-memory architectures.

Block-relaxation methods are generally more advantageous than point relaxation methods in terms of computational/communication density.

Better scalability can of course be gained by extracting further parallelism from the direct stage of the solution. This is however far from being a trivial task, as we shall see in the following, mainly because of the narrow bandwidth of the blocks D_j (5-diagonal in three dimensions, 3-diagonal in two).

In passing, we mention that the parallel solution of block tridiagonal (BT) and scalar pentadiagonal (SP) systems of equations are two of the basic kernels of distributed parallel computing. They are part of the popular NAS benchmarks, as of today one of the most credited benchmark for assessing the performance of commercial parallel architectures [125].

Block algorithms have better parallel efficiency, but are also more CPU demanding per grid point because a direct solver is involved. On the other hand, they generally require less iterations to converge . Hence, the ultimate choice of the parallel scheme must result upon judicious weighting of all of these aspects.

To illustrate this point, let us compare:

1 A point-relaxation three-dimensional partitioning scheme (P cubes , L^3/P points each)

2 A plane-relaxation, one-dimensional partitioning, scheme ($P = L$ planes , L^2 points each).

For case 1), we have

Computation: $6L^2/P$

Communication: $6L^2/P^{2/3}$

and for case 2):

Computation: $4L^2 \cdot L^2$

Communication: $2L^3/P$

Here, the compute density $\chi \sim O(L^2)$ comes from the characteristic complexity of Gaussian elimination, NB^2, where $N = L^2$ and $B \sim L$, whence $\chi \sim O(B^2) \sim L^2$.

The corresponding CCR are therefore given by:

$CCR^{(1)} \sim P^{1/3}/L$

$CCR^{(2)} \sim P/L^3$

We recognize the typical exponents of one-dimensional and three-dimensional partitioning as far as the P dependence is concerned.

The L^{-1} dependence of$CCR^{(1)}$ is familiar for it corresponds exactly to the behaviour of an explicit scheme.

What is new is the factor L^{-3} in $CCR^{(2)}$, which is a clear signature of implicit schemes whose computational density scales like $O(B^2)$ rather than $O(\nu)$, B being the bandwidth of the matrix and ν the number of its non-zero entries.

As a result, on a qualitative ground, block schemes are to be preferred because, first they usually require less iterations, second they offer better communication/computation ratios. Obviously, one should not forget that they're more CPU demanding on a single iteration basis.

In the end, the ultimate choice must necessarily result upon careful evaluation of the interplay between all competing factors.

7.2 Gradient Methods

As already mentioned, gradient methods, and particularly conjugate-gradient (CG) methods, have gained a wide popularity in CFD applications and notably those requiring unstructured grids. This is basically due to their high level of generality and independence of any regularity assumption on the geometry of the problem.

Let us remind the basic computational steps entailed by a CG iteration (cfr. Chap. IV) (without preconditioning).

0.	$s^0 = r^0 = Ax^0 - b$		
For k = 0,1 untilconvergence do			
1.	$y = As$	(mvm)	νN
2.	$\lambda = (r,r)/(s,y)$	(sdot)	$3N$
3.	$\hat{x} = x + \lambda s$	(saxpy)	$2N$
4.	$\hat{r} = r - \lambda y$	(saxpy)	$2N$
5.	$\beta = (\hat{r},\hat{r})/(r,r)$	(sdot)	$3N$
6.	$\hat{s} = \hat{r} + \beta s$	(saxpy)	$2N$
End do			

The computational kernel of the CG iteration consists of the following three basic linear algebra primitives:

one matrix-vector-multiply (*mvm*) ; step 1

two scalar products (*sdot*) ; steps 2, 5

three saxpy ($y = y + s \times x$) (*saxpy*) ; steps 3, 4

The corresponding operation count is (counting *saxpy* as two flop/s): $O_{CG} \sim (\nu + 12)N$, where ν is the degree of sparsity of A (number of nonzero entries per row).

As a first remark, we notice that the CG iteration is rather "chunky", i.e. the compute density $\chi \simeq \nu + 12$ is higher than for other iterative methods, typically featuring $\chi \sim \nu$.

As a second remark, we emphasize the fact that the unstructured nature of the problem only shows up at the level of the *mvm* operation and nowhere else:

$$y_i = \sum_{j \in \eta_i} A_{ij} x_j$$

where η_i represents the neighborhood of node i, usually described by a list of nodes to be accessed by indirect addressing.

The efficiency of this operation may vary significantly from machine to machine depending on whether specialized hardware/software is catered for to deal with indirect addressing (gather/scatter).

On parallel machines, the major issue is **load-balancing** and efficiency of **remote (long-distance) communication** [126].

Let us come to the parallel implementation. Interprocessor communication is required at steps 1, 2 and 3 by *mvm* and *sdot* primitives as well as the convergence check needed at completion of each iteration cycle.

The same steps also demand a corresponding number of synchronization points to proceed with a consistent value of the global parameters λ and β.

The basics of the parallel implementation of *mvm* for arbitrarily patterned matrices have already been discussed in the previous chapter and will not be repeated here.

Parallelization of inner products has also been discussed in the context of convergence checks. Here, the difference is that after all partial contributions from the various processors have been gathered by a "master" processor, the resulting global sum has to be redistributed to all processor (*broadcast*) to compute the value of the global parameters λ, β needed to continue the iteration.

The single CG iteration is a "good" algorithm for parallel computing because of its relatively high compute density. On the other hand, it is potentially exposed to latency issues in view of the relatively high number of synchronization points and global communications (4) per iteration.

For the specific case of the CG algorithm, an issue of primary importance is the **rate of convergence.** As discussed in Chap. IV, this takes us to the subject of **preconditioning,** the key technique to accelerate convergence and reduce the number of iterations.

Let us comment on this point in some detail.

As discussed in Chap IV, preconditioning consists of replacing the original problem $Ax = b$ with an equivalent one

$$\tilde{A}\tilde{x} = \tilde{b} \tag{7.6}$$

where $\tilde{A}$ is as close as possible to the unit matrix I.

In practice one seeks $\tilde{A}$ in the form of a matrix whose eigenvalues are clustered in such a way as to attain the best possible condition number: $\kappa \equiv |\lambda_{min}/\lambda_{max}| \sim 1$.

Manifestly, forcing the transformed matrix $\tilde{A}$ exactly into the unit matrix I is tantamount to solving the original problem by direct substitution.

Just remind the identity:

$$I = L^{-1}AU^{-1} \; if \; A = LU \tag{7.7}$$

In practice one contents himself with approximate, or "incomplete" factorizations, where "incomplete" means that all non-zero elements which would cause deviations of the sparsity pattern of $\tilde{A}$ with respect to A, are suppressed (set to zero) in the course of Gaussian elimination.

Denoting by L_A and U_A the approximate left and right preconditioners, the basic preconditioned CG iteration reads as follows:

0 $\quad r^0 = Ax^0 - b \; ; \; \tilde{r}_0 = L_A^{-1} r_0 \; ; \; s_0 = \tilde{r}_0$
For k = 0,1 untilconvergence do
1 $\quad y = As$
2 $\quad \lambda = (r,r)/(s,y)$
3 $\quad \hat{x} = x + \lambda s$
4 $\quad \hat{r} = r - \lambda y$
5 $\quad \hat{\tilde{r}} = L_A^{-1} \hat{r}$
6 $\quad \beta = (\hat{\tilde{r}}, \hat{r})/(\tilde{r}, r)$
7 $\quad \hat{s} = \hat{\tilde{r}} + \beta s$
End do

This adds a forward sweep (step 5) to the unpreconditioned version.

As we shall see, these steps offer a limited amount of parallel potential even on structured meshes, due to delay-dependencies.

From the point of view of parallel efficiency, the key issue is however a fast method for working out the explicit form of the preconditioners L_A, U_A. This is again not straightforward because it involves the direct solution of a linear system.

At this stage, the reader should not need any further details to convince himself that achieving scalability for CG based schemes is still a challenging task.

Yet, relatively successfull implementations on MIMD architectures (of the order of tens of processors) are reported in the literature [127] and some of them will be described in the next chapter devoted to selected parallel CFD applications.

Direct Methods

Direct methods find widespread use for the solution of CFD problems, not much as stand-alone solvers but rather in combination with iterative solution techniques.

As already discussed in the previous section, one and two-dimensional direct solvers are an essential component of block-iterative methods. In addition, they lie at the heart of preconditioning, a key step for the successfull implementation of gradient-like iterative methods.

Fully direct solvers are generally ruled out for three-dimensional CFD, on account of their unacceptable memory and CPU requirements (cfr. Chap IV). This statement, which is out of question for serial machines, might no longer apply to parallel machines affording sufficient memory and CPU power to accomodate the demand raised by Gaussian elimination methods.

A further motivation for the use of direct solvers is provided by the fact that even the most advanced iterative solvers may run into troubles when the physical complexity makes the matrix A strongly non-symmetric, hence ill-conditioned.

Extracting high parallel efficiency out of direct methods still appears a challenging task [128].

To discuss this point in some detail, let us focus on Gaussian elimination methods.

7.2.1 LU Decomposition

As discussed in Chap. IV, the Gauss elimination technique is tantamount to finding a decomposition of the matrix A into a LU form, where L and U are lower and upper triangular matrices respectively.

Once the LU decomposition is found, the original problem

$$Ax = b$$

is readily solved with a forward sweep $Ly = b$, followed by a corresponding backward sweep $Ux = y$.

Hence, Gauss elimination involves three basic steps:

1. LU decomposition of matrix A
2. Forward substitution $y = L^{-1}b$
3. Backward substitution $x = U^{-1}y$

Step 1 is by far the most demanding in terms of operations count. For a regular matrix of dimension N and bandwidth B, the following operation count results:

$$O_1 \sim O(NB^2); \qquad O_2 = O_3 \sim O(NB)$$

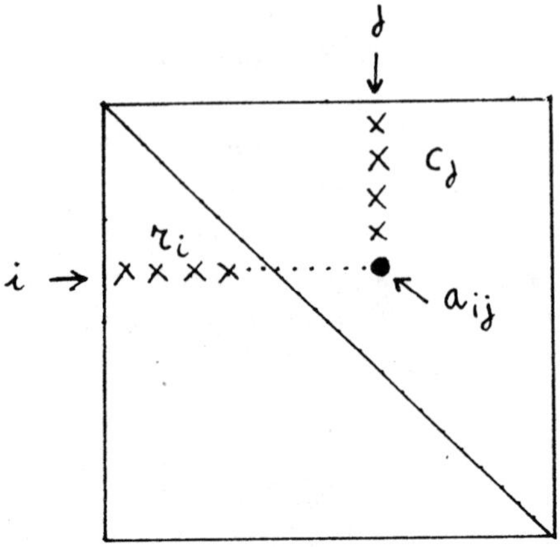

Figure 7.9: The reduction of the element a_{ij} involves the dot product between the row array r_i and the column array c_j

Since $B \sim N^{1-1/D}$ in D dimensions, $O_1/O_{2,3} \sim N^{2-2/D} \gg 1$ for N sufficiently large.

Let us therefore concentrate on the LU decomposition.

Formally, the LU decomposition reads as follows:

$$\begin{aligned} l_{ij} &= (a_{ij} - \textstyle\sum_{k=i-h}^{j-1} l_{ik}u_{kj})/a_{jj} \quad ; i = 1, N \quad , j = 1, 2, ..., i-1 \\ u_{ij} &= (a_{ij} - \textstyle\sum_{k=j-h}^{i-1} l_{ik}u_{kj}) \quad ; j = 1, N \quad , i = 1, 2, ..., j-1 \end{aligned} \tag{7.8}$$

Here h is the half bandwidth of the matrix A: $h = (B-1)/2$. Note that the summations on the r.h.s. of eqs. 7.8 are nothing but the inner product (*sdot*) between the i-th row of matrix L and the j-th column of matrix U (see Fig. 7.9), (summations with negative bounds are implicitly assumed zero).

The difficulties raised by eqs. 7.8 for parallel computing are apparent: **recursion** and **non-locality**.

Recursion is apparent since in order to compute l_{ij} all forerunners, $l_{ij'}$ $j' < j$ are required. This feature has already been discussed with regard to relaxation schemes. Non locality, which is peculiar to direct methods, comes in two flavours: first, the arrays to be dotted, r_i and c_j, do belong to distinct sectors, L and U respectively, of the global matrix A. A second element of non-locality is introduced by pivoting, i.e. the permutation of distinct columns (rows) of matrix A in order to select the maximum a_{jj} for the sake of numerical stability.

A few parallelization strategies can be worked out to do away with these difficulties.

One possibility is offered by **blocking**.

Let us consider the problem of reducing a block of $I \times J$ unknowns lying within the intervals $\mathcal{I} \equiv \{i_1 \leq i \leq i_2\}$ and $\mathcal{J} \equiv \{j_1 \leq j \leq j_2\}$.

It is clear that the inner products between the I row vectors r_i and J column vectors c_j can be executed concurrently by distributing the c_j's across J distinct processors, all sharing the I rows r_i's.

This solution is fine for shared-memory implementations, but is not viable for distributed memory because all processors should own the whole set of row vectors r_i, thus breaking memory scalability.

Consider now the reduction of a diagonal block A_p of size $h \times h$ partitioned into $N_\beta \equiv h^2/\beta^2$ sub-blocks of size $\beta \times \beta$.

A naive distributed memory implementation could proceed as follows: each processor owns a "bunch" of row vectors $\underline{r}_i$ $i \in \mathcal{I}_p = \{i; I_1(p) \leq i \leq I_2(p)\}$ and a bunch of column vectors c_j, $j \in \mathcal{J}_p = \{j; J_1(p) \leq j \leq J_2(p)\}$; $I_2 - I_1 = J_2 - J_1 = \beta$ being the sub-block size.

All "in-block" elements can be reduced in parallel, whereas the reduction of an "off-block" element a_{ij}, $i \in \mathcal{I}_p$, $j \in \mathcal{I}_q$, $p \neq q$, requires $\underline{r}_i$ being sent from processor p to processor q.

The amount of calculation per sub-block is given by $O(h\beta^2)$ corresponding to β^2 inner products between arrays of size h.

The amount of communication is given by $O(2\beta h)$, since each sub-block has to send/receive 2β arrays of size h to all blocks except itself.

The resulting CCR is given by: $O(1/\beta)$, thus highlighting the benefits of blocking.

By identifying N_β with the number of available processors, $P = h^2/\beta^2$, the previous relation reads as $CCR \sim P^{1/2}/h$.

This highlights the fact that small bandwidthes, although convenient for the purpose of sheer serial efficiency, set an additional toll for parallel schemes.

To comply with the recursivity of eq. 7.8, the elements within the sub-block must be reduced according to a well-defined specific temporal sequence of steps.

A typical reducing sequence for a 4×4 block could read as follows (P=4):

1. Processors 2, 3, 4 send c_2, c_3, c_4 to processor 1

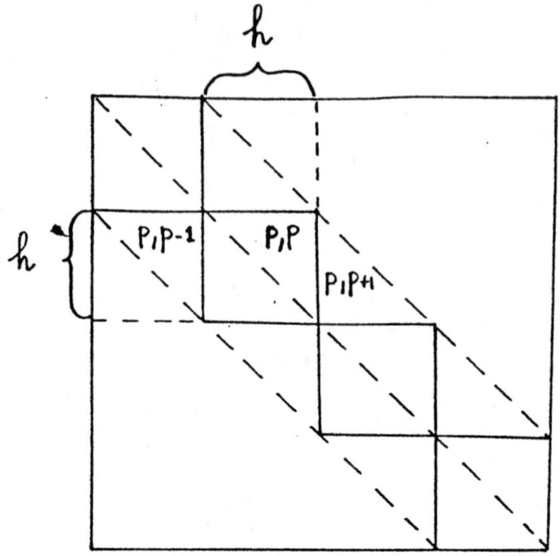

Figure 7.10: Blocking: the block $A_{p,p}$ depends on the previous block $A_{p-1,p-1}$ via the triangular block $A_{p,p-1}$.

2. Processor 1 forms ($l_{11}, l_{21}, l_{31}, l_{41}$).

3. Processor 1 sends r_1 to processors 2, 3, 4

4. Processors 2, 3, 4 form u_{12}, u_{13}, u_{14} in parallel

At this stage, the first row and the first column of the block has been reduced, and the procedure can continue in a similar way to the 2nd column and 2nd row until completion of the sub-block.

This procedure entails P synchronization points. Each block $A_{p,p}$ must be updated sequentially due to the dependence on the previous block $A_{p-1,p-1}$ via the off-diagonal triangular blocks $A_{p,p-1}, A_{p-1,p}$ (see Fig. 7.10).

As a result, the maximum degree of parallelism is upper bounded by the half bandwidth h: $P \ll h$.

Typically, for a three-dimensional problem, one has $h \sim O(L^2)$, $\beta \sim O(L)$ so that $P \sim O(L)$, like for a one-dimensional partitioning.

7.2.2 Parallelizing the backward/forward substitution

The backward/forward substitution steps involve sequential dependence (dependence delay) due to their intrinsic recursivity.

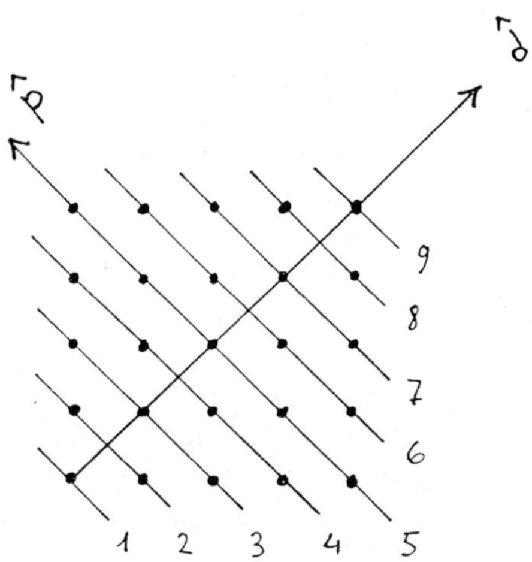

Figure 7.11: Dependency segregation: all points belonging to lines aligned along direction $\hat{p}$ can be updated in parallel. The parallel "front" proceeds along the main-diagonal $\hat{d}$ where dependency makes itself felt.

In one dimension there's no way to escape sequential dependencies other than using approximate iterative (multicolor) techniques. In higher dimensions, $D > 1$, however, "exact" parallelization is still possible along subspaces of dimension $(D-1)$ orthogonal to the one-dimensional characteristics along which dependency takes place. Formally, this is tantamount to a change of coordinates, say $(x, y, z) \rightarrow (\xi, \eta, \zeta)$ such that dependency only affects one coordinate, say ξ, while leaving the other two dependency-free. We may call this technique "dependency segregation".

Consider for clarity a forward step in $D = 2$: the unknown (i, j) only depends on $(i-1, j)$ and $(i, j-1)$, which means that the direction of dependence is the right-handed diagonal $\hat{\imath} + \hat{\jmath}$ ($\hat{\imath}, \hat{\jmath}$ being the unit vectors along x, y respectively).

The result is that all variables belonging to the sequence of left-handed diagonal defined by $i + j = n + 1$, $n = 1, 2L - 1$ can be updated in parallel, one line at a time (see Fig. 7.11).

This procedure is simple - like an explicit method - but has a somewhat limited parallel potential because the number of nodes per line varies from line to line and never exceeds the top value of L points for $n = L$.

More precisely, for a three-dimensional cubic domain of linear size L

(in $D = 3$ the dependency direction is the main diagonal $\hat{\imath} + \hat{\jmath} + \hat{k}$, and parallelization can occur all over the planes perpendicular to this direction), the number of points on plane n, say ν_n is given by [129].

$$\nu_n = \frac{1}{2} \begin{cases} (n+1)n & 1 \leq n \leq L \\ \frac{1}{2}(n+1)n - 3(n-L)(n-L+1) & L+1 \leq n \leq 2L-2 \\ (3L-1-n)(3L-n) & 2L-1 \leq n \leq 3L-2 \end{cases}$$

where each plane is defined by $i + j + k = n + 2$, $n = 1, 3L - 2$.

It is clear that whenever ν_n is not an integer multiple of the number of processors P, a corresponding amount of compute power (i.e. parallel efficiency) is wasted away in the form of load unbalance. This waste is easily quantified as the ratio between ν_n and ν_n^P, i.e. the number of points which would be needed in order for ν_n^P to be an integer multiple of P (obviously $\nu_n^P > \nu$).

Mathematically:

$$\nu_n^P = \begin{cases} \nu_n & if\ mod(\nu_n, P) = 0 \\ P(\frac{\nu_n}{P} + 1) & if\ mod(\nu_n, P) \neq 0 \end{cases}$$

The loss of parallel potential is then measured by:

$$\eta = \sum_{n=1}^{3L-2} \nu_n / \sum_{n=1}^{3L-2} \nu_n^P$$

One can easily check that η drops to about 50 % already for $P \sim \frac{3}{4}L$.

7.2.3 Combining direct and iterative methods

The possibility of using direct methods as a partial step of a global iterative procedure has already been pointed out in the discussion of block-iterative methods. The reverse strategy is also possible: decompose a single direct solver into a sequence of "sub-solvers" coupled together by an iterative loop.

This strategy is especially warranted whenever a set of "loosely-coupled" subgroup of variables can be identified so that "intra-group" variables (tightly coupled) are handled by a direct solver, while "inter-group" loose-coupling is dealt with iteratively.

This strategy, called **substructuring**, in view of its evident links with Computational Mechanics, is particularly interesting in the case of unstructured problems [130].

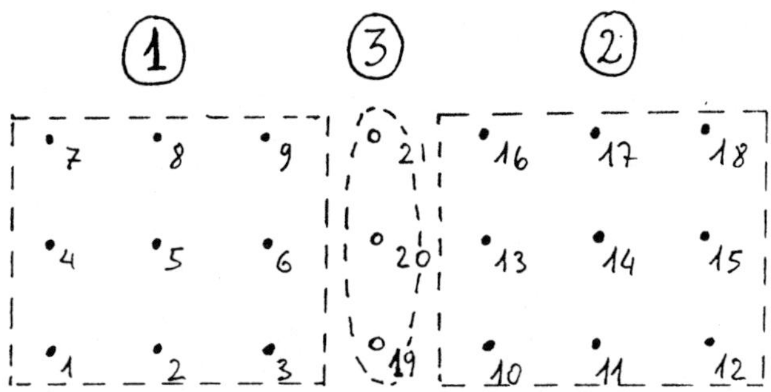

Figure 7.12: Substructuring of a single rectangular domain into three non-overlapping blocks: left (block 1), right (block 2), and middle (block 3).

In principle, however, it can also be used for structured problems, as briefly discussed in the following.

Let us refer to the two-dimensional grid depicted in Fig. 7.12

One possible way to substructure this grid is to partition it into three distinct groups of nodes:

Group 1: Left block made up by nodes 1 to 9

Group 2: Right block made up by nodes 10 to 18

Group 3: Interface nodes 19, 20, 21

With this ordering, the corresponding matrix takes the following *ordered block-diagonal* form (Fig. 7.13):

This form is highly suggestive of a parallel iterative scheme in which the off-diagonal blocks are treated as "perturbations":

$$\sum_j A_{ij}^{(pp)} \hat{x}_j^{(p)} = b_i^{(p)} - \sum_{q \neq p} A_{ij}^{(pq)} x_j^{(q)} \equiv \hat{b}_i \tag{7.9}$$

where caret denotes the actual iteration level, $A_{ij}^{(pp)}$ is the p-th diagonal block, and $A_{ij}^{(pq)}$ the off-diagonal one. This is basically a block-Jacobi

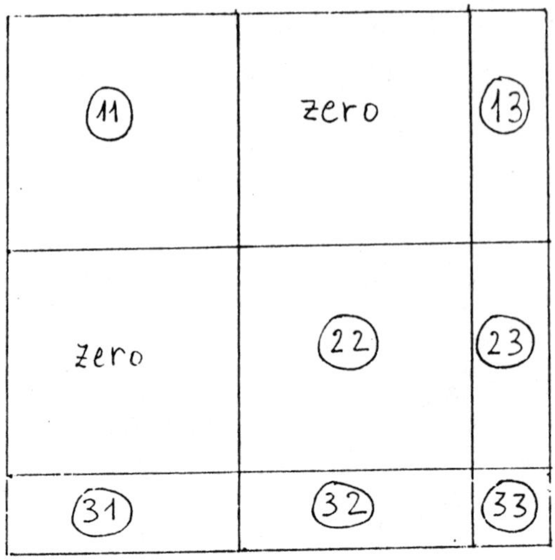

Figure 7.13: The block-ordered matrix resulting from the substructured problem represented in Fig. 7.12

relaxation with block-diagonal ordering. At this stage, one is left with P independent problems which can be solved in parallel with a direct method on each sub-block.

This scheme corresponds to a **non-overlapping** domain decomposition; a certain degree of overlapping may be often desirable to speed-up convergence by a judicious coupling of the blocks [131].

The idea is best illustrated for a one-dimensional system with nearest neighbor coupling giving rise to a tridiagonal matrix (Fig. 7.14,7.15).

In the non-overlapping version, with, say, two processors and 8 grid points, the scheme reads as follows:

Processor 1 computes x_1, x_2, x_3, x_4 using x_5 provided by processor 2

Processor 2 computes x_5, x_6, x_7, x_8 using x_4 provided by processor 1

Since x_5 may not be up to date, this produces an approximation error which is genuinely related to parallelization.

This approximation error is generated at the interface but propagates inside the domains with a high chance of "corrupting" the parallel solution.

With a two grid-points wide interface ($\delta = 2$), the scheme reads as follows:

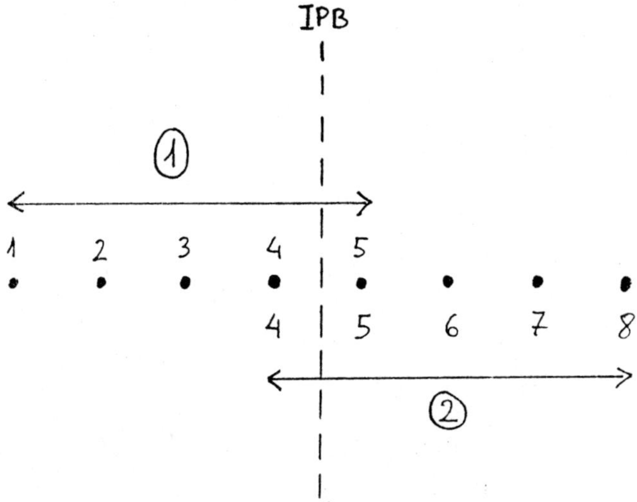

Figure 7.14: Overlapping domain decomposition of a simple one-dimensional grid. Points 4 and 5 are shared among the two processors communicating across the inter-processor boundary (IPB).

Processor 1 computes x_1, x_2, x_3, x_4, x_5 using x_6 from processor 2 and retains only x_1, x_2, x_3, x_4

Processor 2 computes x_4, x_5, x_6, x_7, x_8 using x_3 from processor 1 and retains only x_5, x_6, x_7, x_8

In this case, each processor "wastes" some extra work to compute unknowns which will be subsequently discarded (x_5 for processor 1 and x_4 for processor 2). The advantage is that unlike the non-overlapping case, the error is now originating from x_5, rather than x_4, so that the domain effectively owned by processor 1 is less affected by the approximation error introduced by decoupling the computational domain.

Indeed, it can be shown that if A is diagonal dominant, the "domain decomposition error" decreases with increasing distance from the interprocessor boundary, so that there's a pay-off in "shifting-away" the interprocessor boundary where the error originates from [132], [133]

The trade-off is then quite clear: the wider the overlap, the more we loose efficiency on the single iteration because of redundant calculations. In return, we expect the overall number of iterations be reduced, so that the net outcome is hopefully a win.

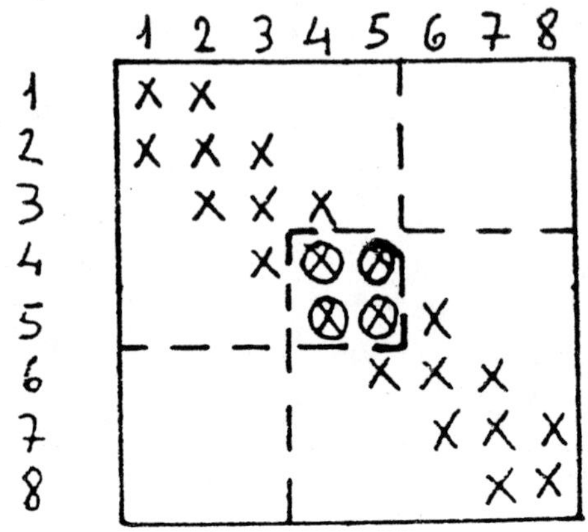

Figure 7.15: The partitioned matrix associated with the overlapped domain-decomposition depicted in Fig. 7.14

7.2.4 Direct solvers for unstructured implicit schemes

Direct methods for implicit unstructured schemes arise mainly in the context of finite-element formulations of fluid flow equations.

As usual, the motivation for finite-elements is geometrical flexibility while that for direct solvers it is eventually lack of convergence of iterative solvers for poorly conditioned matrices engendered by physically complex flows (high-Reynolds, non-Newtonian and the like).

We shall examine basically two categories of methods:

- "Scattered" frontal methods
- Incomplete nested dissection

The efficiency of an unstructured solver is strongly affected by a proper choice of the data structure used to represent the sparsity pattern of the matrix A.

Among others, a highly-prized property is the invariance of the sparsity pattern with respect to the (non-linear) transformations involved by the Gauss elimination procedure.

In this respect, the most popular choice is the so called "skyline" storage pattern. Within this scheme, the matrix A_{ij} is stored as a single one-

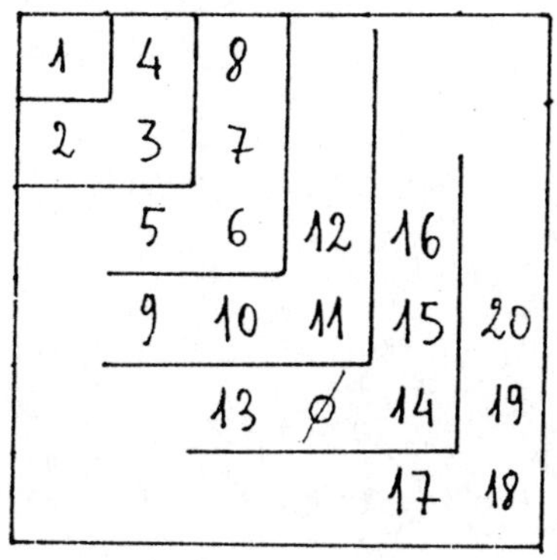

Figure 7.16: The skyline format of a 6×6 sparse matrix

dimensional array, AS, made up of columns and rows of A visited in a sort of "fishbone" ordering (see Fig. 7.16) from top to bottom.

$$AS = a_{11}|a_{21}a_{22}a_{32}|a_{32}a_{33}a_{23}a_{13}|a_{42}a_{43}a_{44}a_{34}|a_{53}0a_{55}a_{45}a_{35}|a_{65}a_{66}$$

Note that zeroes within the skyline pattern must be retained because they fill-in during Gaussian elimination.

The "scattered-box" frontal method

Besides the difficulties already encountered for the structured case, the unstructured solvers raise the usual issue of load-balancing.

In fact, due to the variable profile of the skyline the various scalar products between rows $\underline{r}_i$ of the matrix L and columns $\underline{c}_i$ of the U matrix may entail quite a different number of operations.

From the conceptual point of view, this is similar to the problem of computing in parallel the integral of a "wild" function $w(x)$

$$I = \int_a^b w(x)dx = \sum_{p=1}^{P} \int_{\mathcal{I}_p} w(x)dx; \; \mathcal{I} = (b-a)$$

It is intuitively clear that the chances to get a significant balance between the various sub-integrals I_p are strongly enhanced if each processor

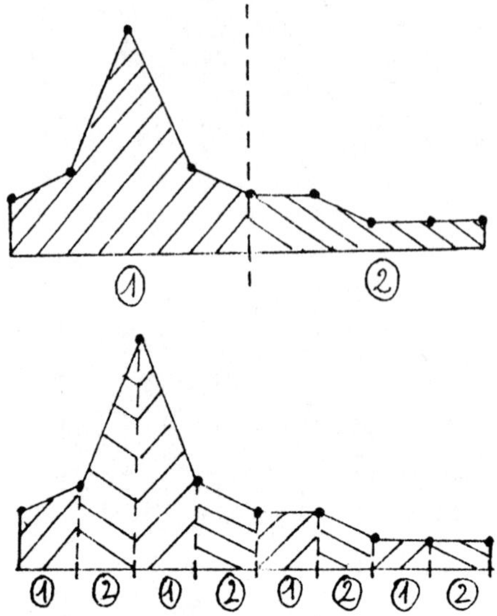

Figure 7.17: Interleaved versus contiguous domain-decomposition for the numerical evaluation of the integral $\int_a^b w(x)dx$

is allowed to probe the entire interval $[a, b]$ by drawing sample points from non-contiguous subintervals.

This points to multicolored ("interleaved") domain-decompositions such as those typified in Fig. 7.17 for the case $P = 2$.

In the context of the LU decomposition, this means that the first column vector is given to processor 1, the second to processor 2, ..., the P-th to processor P and the (P+1)th to processor 1 and so on.

Another technique which is gaining considerable attention is a instance of substructuring known as **incomplete nested dissection.**

7.2.5 Incomplete nested dissection

An incomplete nested dissection ordering of an unstructured mesh is shown in Fig. 7.18.

At the "substructural" level, this mesh is composed by 7 functional blocks, 4 "intra-processor" blocks B_1, B_2, B_3, B_4 interconnected through three inter-processor blocks, or "separators" S_1, S_2, S_3.

The dependence diagram of the incomplete nested dissection can be organized in the form of binary tree (elimination tree) which provides the guideline for the elimination procedure.

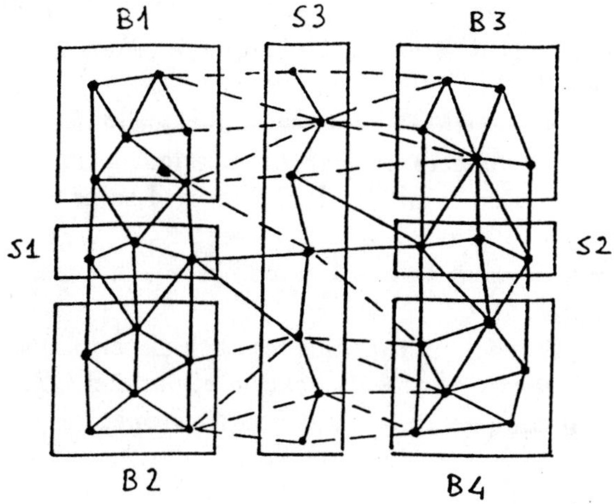

Figure 7.18: Incomplete nested dissection of an unstructured grid

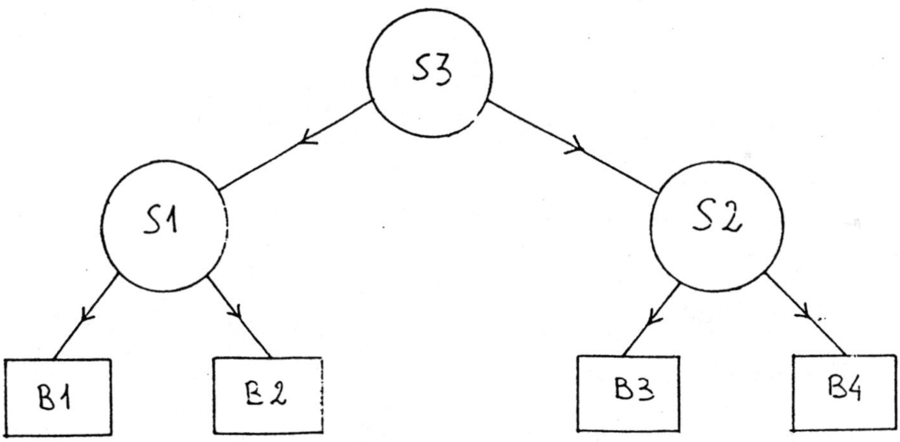

Figure 7.19: The elimination tree associated with the incomplete nested dissection ordering pictured in Fig. 7.18

The elimination tree highlights the potential concurrency: all leaves at the same level can be factorized in parallel. Once blocks B_1 to B_4 are factorized one proceeds up-tree to eliminate the separators S_1 and S_2 also in parallel, and finally S_3.

It goes without saying that at each level of the elimination tree one must pursue the removal of unnecessary synchronization, usual objectives of load balancing, surface/volume minimization, discussed several times throughout this book.

Advanced finite-element to block mappings which pursue these goals are described in [110].

We conclude by remarking that this research direction looks very promising especially with regard to the application of parallel CFD to problems of significant physical complexity, such as non-Newtonian flows, polymer flows and the like.

7.3 A pause of reflection

At completion of an intense section of this book, we feel it appropriate to take one step back and spend some words of comment.

Let us remind the logical set of options examined so far for grid methods

	Structured	Unstructured	
Explicit	ES	EU	
Implicit	ISI	IUI	Iterative
	ISD	IUD	Direct

which split into six "logical" classes of methods depending on three dichotomic options:

1. Explicit / Implicit
2. Structured / Unstructured
3. Iterative / Direct (Applies only to implicit) [1]

Let us discuss the three options separately.

[1] For finite-elements we know that the Direct/Iterative option would set up also for explicit methods because the mass matrix is not diagonal. However, we shall assume that the mass matrix is lumped so that this option only applies to implicit methods.

7.3.1 Explicit / Implicit

Explicit methods parallelize in a fairly transparent and straightforward fashion. This comes to no surprise if we recall that they represent the closest approximation to what Nature does: local and causal connections in space-time.

On a practical ground, their scalability is hampered by a generally poor compute/communication ratio (CCR), which tends to one (one flop/word) in the scaling limit $P \rightarrow L^3$ (one processor per grid point).

Such a limit is fairly too low for present-day computer technology which can hardly afford CCR's below the range of $\simeq 10$ Flops/word. As a result explicit methods are better suited to (relatively) coarse-grain implementations.

Implicit methods do not parallelize transparently, and generally require a great deal of ingenuity before significant parallel potential can be extracted out of them.

This is the price to pay for committing the crime of breaking causality, hence bringing non-locality into the schemes.

Non-locality can either be handled head-on by direct methods, or partially ruled out by iterative methods, which restore causality along the "relaxation-time" direction.

Very often, a judicious combination of direct and iterative techniques yields the best pay-off.

7.3.2 Iterative/Direct

Iterative methods -like the explicit ones- suffer of generally poor CCR's. Not so for direct methods which, once successfully parallelized, support significantly higher CCR's. As an order of magnitude estimate, the gain factor is about $B^2/\nu N_{it} \leq B/\nu$, where B is the matrix bandwidth, ν the number of non-zero entries and N_{it} the number of inner iterations per time step required to converge.

The assumption $N_{it} < B$ is a fairly loose one, and should be checked out on a case-by-case basis.

7.3.3 Structured/Unstructured

The issue of structured versus unstructured mesh (for complex geometries) is still a matter of debate in the CFD community.

The advantages of structured mesh are clear: simple data structures, direct addressing, no need for pointers to know "who is who" in the mesh, are for some researches just too good to be given up with.

This simplicity carries over into the realm of parallel computing: transparent load-balancing being the primary reward.

According to the other school of thought, no structured mesh, no matter how clever, can compete in terms of grid-point requirements with an unstructured mesh for truly complex geometries.

Hardware-wise, unstructured computations are generally well assisted by modern architectures: gather/scatter primitives for indirect addressing, and non-metric switching networks being just two of the most relevant features.

As of today, it appears that load-balancing, both in the computation and in the communication stages, is the main issue to be cleared in order to raise the competivity of parallel unstructured solvers.

At this stage, it's worth mentioning a third -emerging- alternative, known as **block structured** mesh, which should retain the advantages of both procedures while leaving aside the corresponding pitfalls. The idea is to represent complex geometries as a collection of blocks each of which is meshed with its own structured grid. This provides a substantial simplification of the in-block computations, while leaving the burden of geometrical fidelity to appropriate "inter-block patching" interfaces. Parallel multiblock schemes have met a significant success especially for the external flow computations and are also making their way into commercial packages [134],[135].

7.4 Spectral Methods

As to spectral methods, their efficiency is behindered by global communication requirements, which result in $CCR \sim O(1)$.

Very often SM can boast excellent single-processor performance (sustained speeds close to the machine peak), which keeps the method fairly competitive over a significant range of processors.

7.5 Particle Methods

Local synchronous PM, like Lattice Gas and Lattice Boltzmann, are basically in the class of explicitly structured grid methods, with the notable exception that they feature much higher compute density, hence better scalability. The question is up to which extent this extra-computation is really valuable for fluid-dynamic purposes.

Local asynchronous PM, like MD, bear deep similarities with unstructured explicit schemes with a time-varying connectivity. Again, they feature a higher computational density and consequently a better parallel efficiency as long as load-unbalancing issues can be kept under control.

Global asynchronous PM, like the Vortex Method, are affected by global computation and communication at a time. The resulting CCR is $O(P/N)$, in the most naive version, which is pretty good for parallel efficiency. The problem is to keep scalability also for sophisticated schemes characterized by local computation requirements.

Chapter 8

Case studies in parallel CFD

In this final chapter, we shall present a representative sample of selected parallel CFD applications.

This sample is by no means exhaustive of the huge set of publications devoted to this rapidly expanding subject, nor does it reflect any judgement of quality by the authors.

The examples dicussed in this chapter are the following:

1. Rayleigh-Bénard turbulent convection with Lattice Boltzmann (on IBM SP1)

2. Industrial CFD applications with Parallel Unstructured Conjugate Gradient solver (on n-CUBE, IBM SP2, Cray Y-MP)

3. Three-dimensional homogeneous incompressible turbulence with parallel spectral code (on CM-2)

4. Incompressible flow past solid bodies with Parallel Vortex Methods (on iPSC/860)

5. Parallel Rarefied Gas Dynamics (on CM-5)

8.1 Lattice Boltzmann on the IBM SP1

The Lattice Boltzmann (LB) method belongs to the simplest class of parallel CFD algorithms: explicit and uniform in space. As previously discussed, owing to its ties with kinetic theory, LB affords quite a favourable communication to computation ratio.

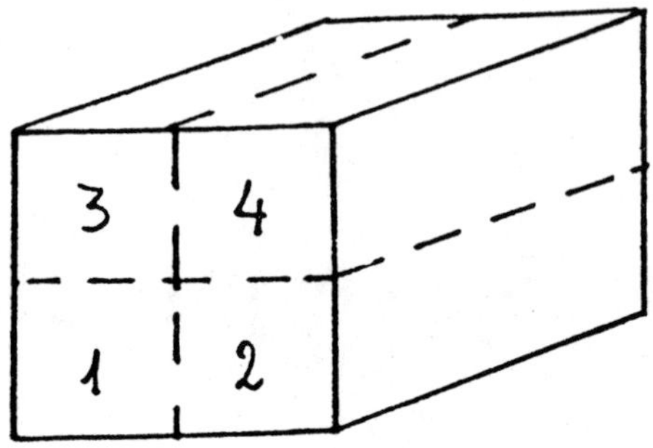

Figure 8.1: Two-dimensional lexicographic domain decomposition of the three-dimensional box

As a such, it is a good parallel performer on virtually any parallel platform. Here we shall focus on a specific implementation on the IBM SP1 scalable parallel computer (See Appendix for architectural details) with up to 128 processors [136], [137].

The implementation on SP1 is based upon a two-dimensional domain decomposition; this choice limits its scalability in the highly scalable regime ($P \sim L^3$) but is well suited to the actual granularity of the configuration in point ($P \leq 128$).

The global domain is a box consisting of $N_x \times N_y \times N_z$ grid points and the subdomain-to -processor assignement is based on a simple two-dimensional lexicographic partitioning (See Fig. 8.1)

Within this domain-decomposition, the collision step is embarassingly parallel, while the move-step requires nearest-neighbor communication along the 18 directions defined by the discrete speeds $\vec{c_i}$ of the FCHC lattice (Fig. 4.6).

In the move phase each processor must communicate the outgoing/ingoing populations f_i flying across the subdomain boundaries.

To enforce full alignement with the serial version, each processor domain is "augmented" with a halo of cells which permits to handle interprocessor data exchange at the same guise as normal boundary conditions.

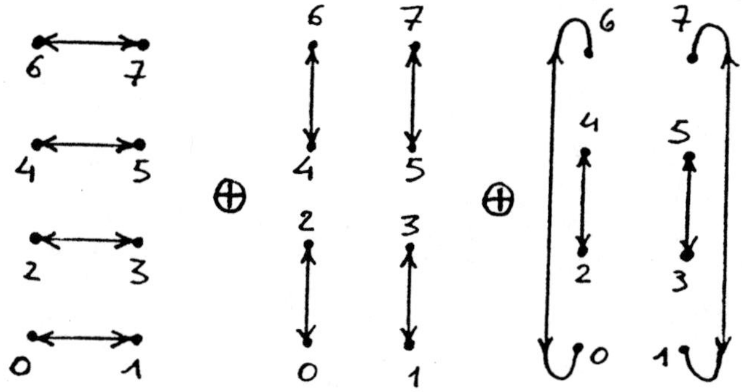

Figure 8.2: Contention-free communication pattern for horizontal (left) and vertical (center and right) data exchange on a 8-processor configuration

The interprocessor communication pattern was designed with the following three criteria in mind:

1. Promote the highest degree of concurrency between different processors during the communication stage, to take full advantage of the SP1 switching network
2. Minimize the network buffering activity due to messages hanging behind the scene, i.e. send/receive operations awaiting for the corresponding receive/sends.
3. Hide machine latency

Due to the regularity of the application, goal 1) can be achieved by organizing the communication pattern in terms of $P/2$ disjoint pairs of sender/receivers ("communication dimers"). This pattern is just perfect to exploit the switch capability to let any set of disjoint pairs of processors communicate simultaneously across the network with no contention.

Goal 2) is easily achieved by taking advantage of the MIMD nature of the SP1 machine, and letting different processors go through different execution paths of the same program (Single Program Multiple Data, or SPMD mode).

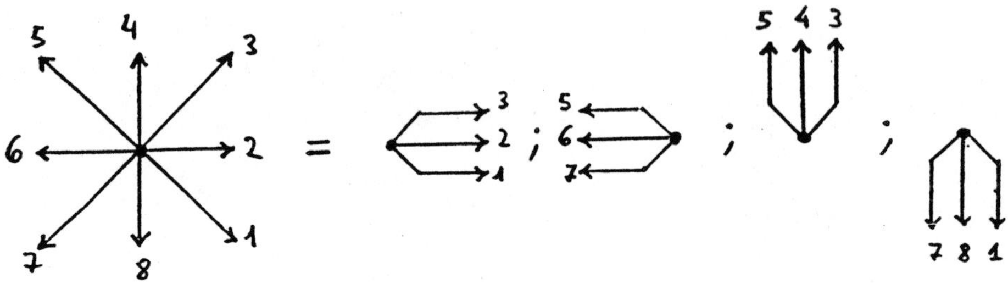

Figure 8.3: Message coalescence: all data propagating along the same direction are collapsed into a single message

In practice, this means that if the "sender" processor is executing a stream of message passing instructions such as:

send ( )
receive ( )

its "receiver" partner should execute in reverse

receive ( )
send ( )

This promotes the formation of "communication bonds" which minimize the occurrence of undefined waiting-states causing high buffering activity, hence anomalous loss of efficiency.

Owing again to the regularity of the application, all communication bonds support the same amount of data traffic, so that load-unbalance exposures are virtually absent.

Finally, objective 3) suggests messages be coalesced as much as possible. This is easily achieved by merging data flowing along the same direction into a single message.

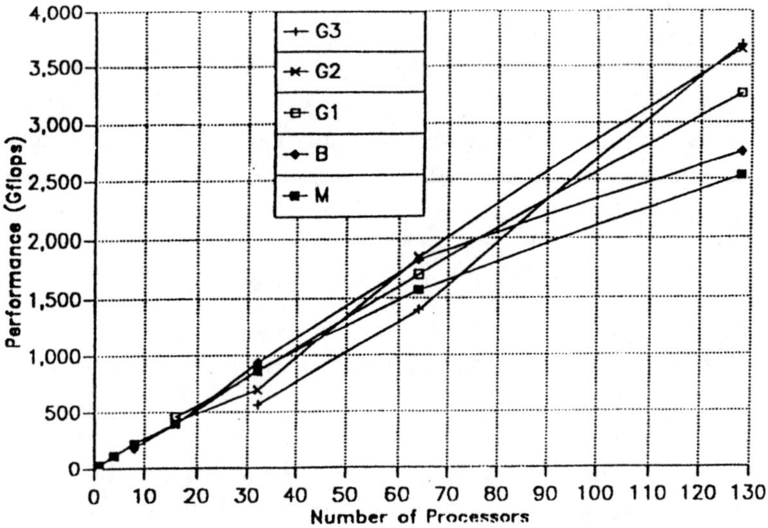

Figure 8.4: Raw performance (Gflop/s) of the cases (M) to (G3) as a function of the number of processors

The parallel LBE code has been implemented on up to 128 processors, for a series of different resolutions:

	$N_x \times N_y \times N_z$	Memory (MB)
M ("medium")	64 × 96 × 128	72
B ("big")	128 × 192 × 256	576
G1 ("giga1")	256 × 192 × 256	1152
G2 ("giga2")	256 × 256 × 256	1536
G3 ("giga3")	256 × 512 × 256	3072

The results are shown in Fig. 8.4

These results show that all of the present jobs enter the Gflop regime between 32 and 48 processors, i.e. at a sustained to peak performance ratio (STPP) between 15÷25 percent (peak-performance with 32 processors being 4 Gflop/s).

Large jobs, while outperformed by smaller ones in the low-granularity region ($P \sim 32$), do recover in the high-end processor region ($P \sim 128$) by attaining almost 4 Gflop/s with 128 processors. This is a typical behaviour of hierarchical memory architectures. By increasing the number of processors, at a given job-size, each processor owns less data so that data access via high-speed buffers ("cache") becomes more efficient.

It is worth noting that the STPP at P = 128 is almost the same as on the single processor.

These results put an incentive on improving the superscalar features of the serial code, while witnessing at the same time an almost perfect signature of scalability.

The present LB code has been used for studying the statistical properties of fully developed thermal turbulence and Rayleigh-Bénard convection [138].

8.2 Engine combustion with unstructured conjugate gradient solver

Parallel versions of unstructured solvers based upon preconditioned Conjugate-Gradient-like methods appear to be the most popular choice for the parallelization of CFD codes dealing with complex geometries such as those commonly encountered in the industrial context.

The example presented in this section hereafter is drawn from a recent paper by Bachler and Greimel [139], who discuss one such implementation on the nCUBE2 multicomputer for real-life applications arising from the automotive industrial context.

The linear system of equations arising from a finite-volume discretization of the Navier-Stokes equations is handled with a Krylov subspace method (ORTHOMIN) and incomplete LU factorization. Four partitioning strategies are examined: data decomposition (DD), coordinate bisection (CB), recursive coordinate bisection (RCB) and recursive spectral bisection (RSB).

The parallel code gives rise to two basic types of communication:

1. Local point-to-point

2. Global

Local communication is triggered whenever information has to traverse the interprocessor boundary (IPB). To this purpose, IPB's are augmented with "halo cells" serving as send/receive temporary buffers.

Global communication arises within the linear solver whenever dot products are required, mainly to enforce orthogonality of the residual vectors within the inner iteration loop. Global communication is also required at completion of each inner and outer iteration loop before entering the convergence check.

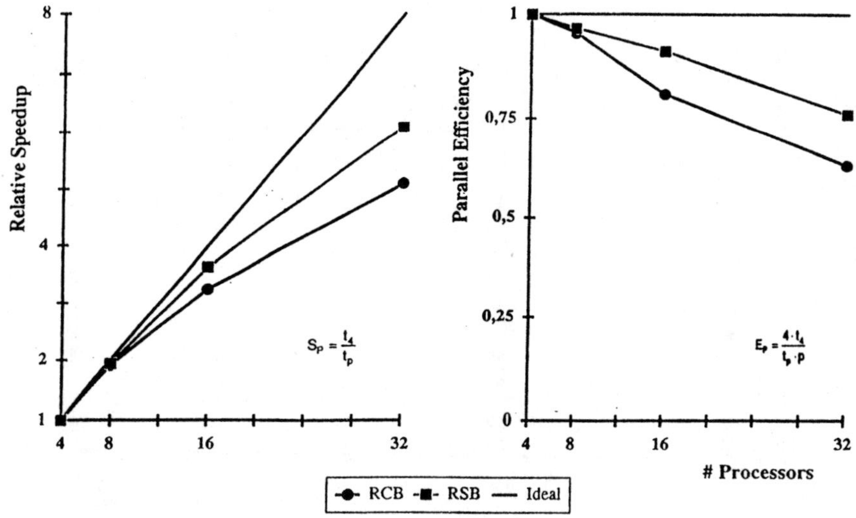

Figure 8.5: Speedup and efficiency for the Intake Manifold using RCB and RSB on the nCUBE

Since the nCUBE architecture caters for non-blocking I/O (the CPU can proceed concurrently with I/O operations), whenever possible each processor first updates the internal layer next to the IPB, then sends it to the neighboring processor while proceeding at the same time with the updating of the next inner layer.

The speedup for an intake manifold consisting of about 30K cells using RCB and RSB are shown in Fig. 8.5 (the benchmark operates at constant number of inner and outer iterations).

From this figure we see that RSB outperforms RCB by virtue of a better surface/volume ratio. However, even RSB shows signs of saturation already with 32 processors. This is due to relatively small size of the problem.

Significant improvements can be obtained by moving to larger job sizes and measuring the **scaled speedup**, i.e. the speedup obtained by raising the size of the problem in proportion to the number of processors used in the computation.

The scaled speedup for a cooling-jacket series of runs (about 300k cells) are presented in Fig. 8.6

In this figure the execution time T is normalized to $T_{nCUBE}(1,1)$, the time taken by a single nCUBE processor on a 1k cell problem. Obviously, the straight line $T = T_{nCUBE}(1,1)$ represents ideal speedup and linear com-

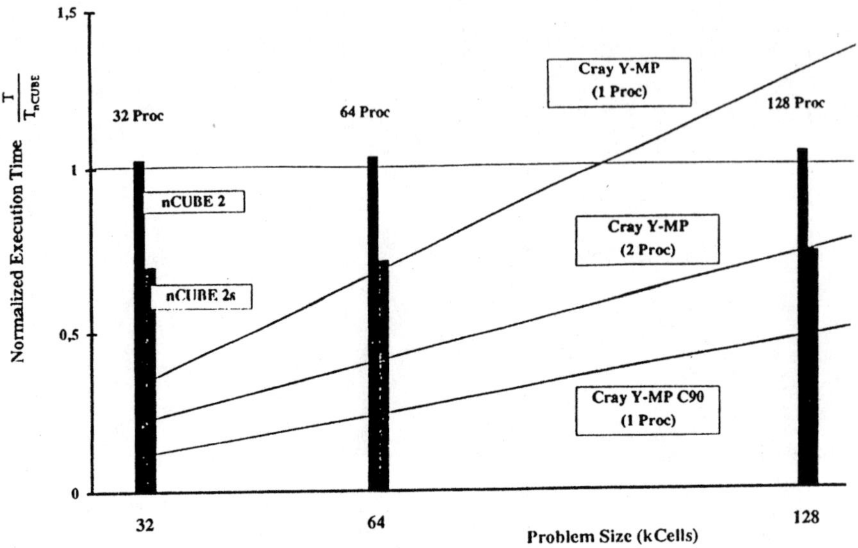

Figure 8.6: Scaled speedup for the Cooling Jacket

putational complexity $O(N)$.

From this figure we see that by running appropriately large problems, scaled speedup can be maintaned up to 128 processors with no apparent signs of degradation.

Also quite interesting is the comparison with powerful vector supercomputers such as Cray Y-MP. Here the main question is how many nCUBE nodes are needed to match the performance of a single Cray Y-MP processor.

From Fig. 8.6 we infer about 100 for the Y-MP and 800 for C-90. These figures reflect rather closely the respective peak performances (310 Mflop/s for 128-way nCUBE, 333 Mflop/s for Y-MP and 900 for the C-90).

In order to preserve scalability over a wide range of processors, the nCUBE provides a fairly balanced architectural CCR, i.e. 2.2 MB/s versus 2.4 Mflop/s. This results, however, in a rather limited raw performance because the single node is merely too slow.

This point is neatly highlighted by comparing the nCUBE performance with the performance obtained on a medium-grain machine such as SP2, affording huge compute power per single node.

The speedup obtained on a SP2 with up to 32 processors running the cooling jacket benchmark with about 300k cells is shown in Fig. 8.7

Although scalability cannot be preserved beyond $O(10)$ processors (mostly

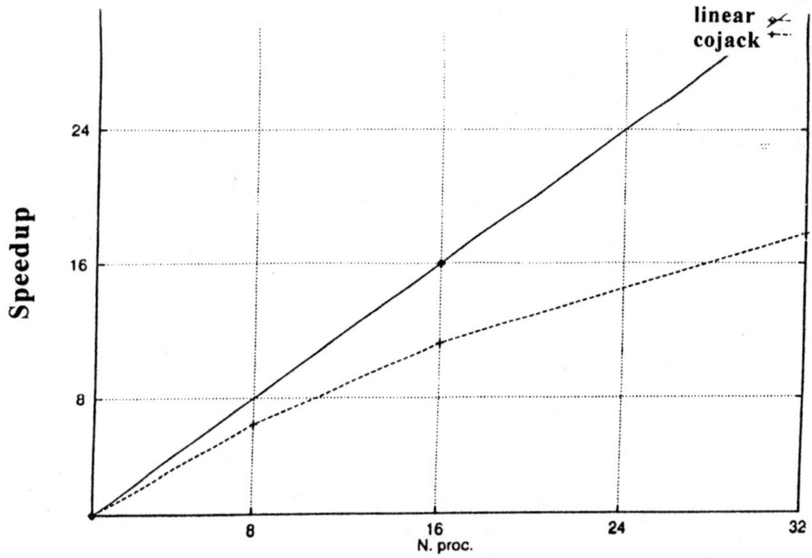

Figure 8.7: Speedup for the Cooling Jacket on the SP2

because of non optimal implementation of global communication), a 16 processor configuration yields a raw performance of about 1 Gflop, more than 3 times a Cray Y-MP, even if the speedup is only about ten [139]. Significantly higher performance can be anticipated by resorting to tree-implementation of global communication on one side [140], and to optimal domain-decomposition on the other.

8.3 Parallel Spectral Codes for Homogeneous Incompressible Turbulence on the CM-2

Spectral methods are the mainstay for the direct simulation of fluid turbulence. The main limitation of these simulations is computer memory which bounds most single-processor simulations to resolutions of about 128^3.

In a recent paper, S.Chen and X.Shan [141] present a detailed study of fluid turbulence with up to 512^3 grid-points.

The calculation was performed on a CM-2 machine with up to 64k (65536) bit-serial processors, 8 GB of core memory and 20 GB disk memory (data vaults). The computational kernel of the application is the Fast-Fourier Transform (FFT).

Parallelization takes place along $X = const$ planes in a typical one-dimensional partitioning set up, and is achieved partly with (CM-SSL) library routines and partly with user-written code.

The language is CM Fortran, basically Fortran 77 extended with data parallel constructs written in Fortran 90. This makes the code quite readable and intuitive. For instance, the relation between velocity $\vec{v}$ and vorticity $\vec{\omega} = \vec{\nabla} \times \vec{v}$ simply reads:

$$\omega_x = (0., 1.) \cdot (k_y \cdot v_z - k_z \cdot v_y)$$

$$\omega_y = (0., 1.) \cdot (k_z \cdot v_x - k_x \cdot v_z)$$

$$\omega_z = (0., 1.) \cdot (k_x \cdot v_y - k_y \cdot v_x)$$

where k_x, k_y, k_z are real arrays representing the coordinates in Fourier space. This is a typical "loop-less" form which makes SIMD data-parallel programming so close to standard mathematical notation.

A single $256 \times 512 \times 512$ FFT takes 8 s/step and since there are nine FFT's per step, the code proceeds at a pace of 72 s/step.

This figure is almost doubled when passing from single to double-precision (actually, as expected, the authors find that single precision is good enough to any practical purpose).

Performance data are reported for three machine configurations: quarter, half and full, and four problem sizes: 64^3, 128^3, 256^3, 512^3. The largest problem requires 5.9 GB of core memory. The results are summarized in the table below (elapsed time per step).

	64^3	128^3	256^3	512^3
1/4	0.611	4.564	36.676	
1/2	0.332	2.357	18.573	
1	0.332	1.241	9.532	76.69

Table 8.1 - Elapsed time of the spectral code as a function of the grid-size and machine granularity

From these data, we observe that good scalability is achieved for all but the smallest problem sizes.

Raw performance for the largest job is reported at 2.2 Gflop/s. Although this is only 8 % of the peak performance (27 nominal Gflop/s), the present problem runs 30 % faster than on a 4-way CRAY-2 vector multiprocessor (peak speed around 2 Gflop/s).

This parallel CFD code has been used to gain insight into the scaling laws of homogeneous turbulence.

8.4 Incompressible flow past solid bodies with Parallel Vortex Methods

Vortex methods offer an appealing pathway to the numerical simulation of high Reynolds flows in complex geometries.

As we have seen in chapter IV, there are two basic strategies to compute the long-range interaction between vortices: Direct (DST) and Fast (FST) summation techniques. An interesting comparison between the parallel implementation of DST and FST summation techniques can be found in [142], for the case of single and multiple vortex rings on the Intel iPSC/860 parallel computer. The Intel iPSC/860 was an hypercube machine which has recently been replaced by the Intel Paragon XP (see Appendix A).

The parallel DST scheme is based on a one-dimensional ring topology, and consists of the following five steps (for each processor):

1. Update the state of resident vortices (vortices assigned to that given processor)

2. Copy resident vortices (host copy) into temporary data (guest copy)

3. Send guest data to front neighbor processor and receive guest copy from the rear neighbor

4. Update resident host copy according to the interaction with guest copy

5. Iterate steps 3) and 4) until completion of all possible interactions, i.e. $P/2$ times for P processors.

As discussed in chapter IV, within this arrangement, all the resident vortex information has to be passed on to all other processors. This results in a high communication burden, which is, on the other hand, compensated by the computational load, scaling like N^2.

The final result is outstanding parallel efficiency but unacceptable computational complexity.

This motivates the analysis of parallel efficiency of FST. To this purpose, the authors consider the same ring-partitioning adopted for DST, except that now each subdomain can further be split into a set of sub-cells, or boxes.

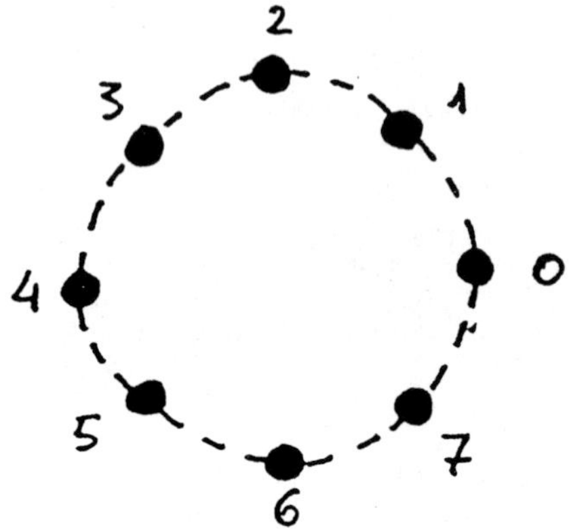

Figure 8.8: Ring topology for a 8-processor implementation of the parallel vortex method

Within each processor, vortex elements are clustered into groups according to the boxes they lie within, and a multipole expansion computed about the center cell of each group.

This arrangement generates four classes of interactions (Fig. 8.9)

In-processor near-field (IN)

In-processor far-field (IF)

Off-processor near-field (ON)

Off-processor far-field (OF)

According to Fig. 8.9 Cells a-b, a-c, c-d, c-e are representative examples of IN, IF, ON, OF interaction respectively.

"IN" interactions are treated via direct summation, while for "IF" interactions multipole expansions are used. Of course, none of the two requires interprocessor communication. "ON" interactions are also treated by direct summation and imply the exchange of all interface vortex elements. Finally "OF" pairs are handled by fast multipole summation, and involve the exchange of the resident multipole expansion coefficient from one processor to

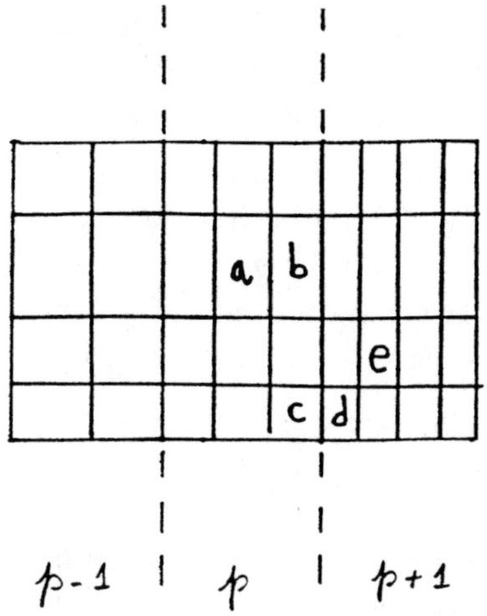

Figure 8.9: Different types of interactions with Fast Summation Techniques

all others. This long-range data exchange is handled by means of the same ring-topology adopted for the N^2 direct method.

Typical results reported for $N = 10^4$ vortices on 32 processors of a iPSC/860 parallel machine are:

$$t_{cal}^{DST} \sim 0.75s;\ t_{com}^{DST} \sim 0.0015s; t_{cal}^{FST} \sim 0.05s;\ t_{com}^{FST} \sim 0.0004s$$

From these figures we deduce that FST are largely preferable, because they can achieve about a factor 15 reduction of the computation time while keeping the communication costs below one percent of the computational cost. A word of caution is however appropriate.

This is somehow the easiest instance of FST, in the sense that the vortex distribution is uniform, so that no adaptive tree-code techniques are required, and the geometry simple enough to rule out any issue related to the handling of boundary layers. A comprehensive discussion of these two issues goes way beyond the scope of this section and can be found in the specialized literature [82] and references therein.

8.5 Parallel Rarefied Gas Dynamics

We conclude this series of sample applications with an example drawn from the context of rarefied gas dynamics. The starting point is a model Boltzmann equation, known as BGK after Bhatnagar, Gross and Krook [143], in which the complex non-linear collision operator is replaced by a much simpler relaxation about a local equilibrium distribution function.

The BGK equation takes the following form:

$$\frac{\partial f}{\partial t} + \vec{v} \cdot \frac{\partial f}{\partial x} = \lambda(f - f^{eq}); \; f = f(\vec{x}, \vec{v}, t) \tag{8.1}$$

where λ is the inverse relaxation time and f^{eq} the local equilibrium, taken in the form of a local maxwellian:

$$f^{eq} = \frac{\rho e^{-(v-u)^2/(2T)}}{(2\pi T)^{3/2}} \tag{8.2}$$

Here $\rho, \vec{u}, T$ are the fluid mass, speed and temperature, and result upon direct integration over the velocity space, according to:

$$\begin{aligned} \rho &= \textstyle\int f d\vec{v} \\ \vec{u} &= \frac{1}{\rho} \textstyle\int \vec{v} f d\vec{v} \\ \rho T &= \textstyle\int \frac{1}{2}(\vec{v} - \vec{u})^2) f d\vec{v} \end{aligned} \tag{8.3}$$

Conservation of mass, momentum and energy must be secured by imposing that f^{eq} carries exactly the same density, velocity and temperature as the non-equilibrium distribution f:

$$\int (f - f^{eq}) d\vec{v} = 0 \; ; \int \vec{v}(f - f^{eq}) d\vec{v} = 0 \; ; \int v^2(f - f^{eq}) d\vec{v} = 0$$

The numerical algorithm to solve eq. 8.1 proceeds in 4 steps [144]:

1. Upwind differences in space
2. Runge-Kutta time-marching
3. Gaussian quadrature for the velocity space
4. Least-squares method for conservation

The parallel implementation of this algorithm is based upon a data decomposition method across the spatial dimension. This choice is very natural one if we consider that most of the computational work is to be performed in the collision, steps 3) and 4). With a spatial decomposition, this work can

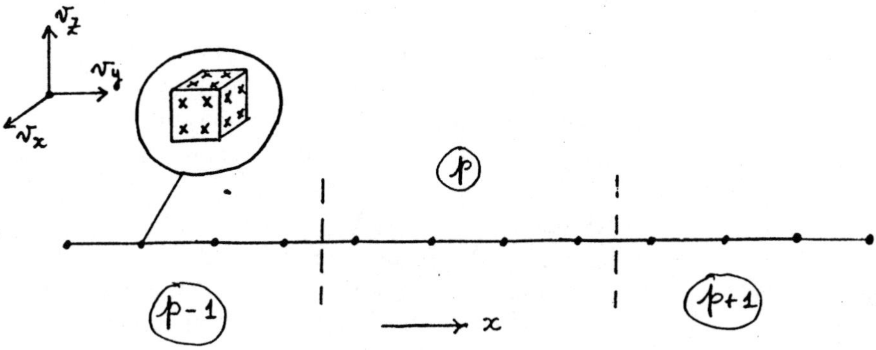

Figure 8.10: One-dimensional domain decomposition of the spatial grid. Each grid point is associated with a corresponding three-dimensional set of discrete speeds

be completely performed in-processor with no need of data communication (embarassingly parallel).

Data communication, like in any explicit time-marching scheme, only occurs at spatial boundaries. This is relatively little information, given the fact that only first order spatial derivatives are involved. Fig. 8.10 illustrates the strategy for the actual case of one space dimension and three velocity dimensions:

With this arrangement, the parallel algorithm takes the neat connotation of a structured explicit scheme, but one with an outstanding favourable communication to computation ratio.

This all traces back to the amount of work to be performed locally in velocity space. Assuming NG gaussian quadrature points per cublet, it is readily checked that step 3) scales like NG, while step 4) scales like NG^2.

Let W_k be the array of Gaussian weights used for quadrature purposes, the discrete density, say at point l, is given by:

$$\rho_l = \sum_{k=1}^{NG} W_k f_{kl}$$

where k stands for a three-dimensional index in velocity space.

On the other hand, the discretized version of step 4) can be shown to take the form:

$$(f_k^{eq} - f_k) = \{(A^T[AA^T]^{-1}A) - I\}(f_k - f_k^M) \tag{8.4}$$

where I is the identity, A is the $(3 \times NG)$ matrix made up with the gaussian weights and f_k^M is the numerical maxwellian corresponding to the numerical density, velocity and temperature obtained via quadrature of eqs. 8.3

Note that since A is a constant matrix, one can define an auxiliary matrix $\tilde{I} \equiv A^T(AA^T)^{-1}A$ and compute it before-hand just once per run. As a result, the expression 8.4 can be completed in $O(9NG^2)$ operations.

It's also worth nothing that the factor 3 in front of NG comes from the fact that in one spatial dimension, only three hydrodynamic quantities, ρ, u_x, T are required. As a result, moving to three spatial dimensions, the factor 3 becomes 5, so that the computational complexity is less than three times higher, 25 against 9.

Summarizing, the parallel BGK algorithm has a compute density scaling like $O(NG^2)$ and a communication density scaling like $O(NG)$. The result is a net factor $O(NG)$ gain with respect to "conventional" hydrodynamic schemes working in configuration space only.

As to the actual numerical value a natural choice is $NG = 7^3$ which ensures exact integration of 13-th order polynomials thus complying with an important method of kinetic theory known as Grad's thirteen moment method.

On the practical ground, this means $NG = 343$ and achieves therefore a two-order of magnitude improvement with respect to the CCR of hydrodynamic explicit methods.

With this preparation, it is no wonder that a highly optimized version of this parallel BGK code was able to achieve as much as 61 Gflop/s on a 1024-node CM-5 machine, i.e. half of the theoretical peak.

A synthetic view of this impressive application are reported in the following table, drawn from the original reference:

N of processors: 1024

N. of grid points: $512 \times 1024 = 524288$

N. of unknowns: $179,830,784$

N of flop/s per step: $4.76 \cdot 10^{10}$

CPU seconds: 0.784

Sustained speed: 60.71Gflop/s

Percentage of peak speed: 46%

At this point, the reader might have noticed many similarities between this sample application and the Lattice Boltzmann application presented earlier on. And indeed similarities are there. The major difference, besides the details of the time marching schemes, is that the computational stencil in velocity space of the LB method is a very peculiar one, for it must comply with special symmetry requirements in order to recover the Navier-Stokes equations.

This sets strong constraints on its geometrical flexibility. On the other hand, since only the hydrodynamic regime is to be recovered, LB needs to carry much less information from velocity space than BGK: 18 discrete speeds suffice in 3D, as opposed to the $7^3 = 343$ speeds of BGK.

To sum up, LB is somehow the "minimal" kinetic-theory-based CFD scheme, lying in between BGK and "classical" purely-hydrodynamic explicit methods.

Whether the outstanding potential for parallel computing offered by kinetic theory can make a major impact on CFD remains an open research issue.

Appendix A

Parallel computer architectures

In this appendix, some parallel computer architectures are sketched.

A.1 The IBM SP1/SP2 architecture

The IBM SP1 is a distributed memory MIMD architecture, and represents the entry point in the family of IBM Scalable Parallel machines. This family has been designed with the intent of leveraging the outstanding IBM's RISC technology and combine it with high speed networking for interprocessor communication.

Each node of the SP1 is a powerful RISC workstation with independent processing and I/O capabilities.

The pool of nodes are racked into "frames" and communicate via a fast switch able to support up to thousands of simultaneous disjoint pairs of communicating nodes.

Distinctive features of the SP1 are:

Machine type: RISC-based distributed-memory multi-processor

Model: IBM 9076 SP1

Clock cycle (ns)	16
Theor. peak performance	
per proc. (Mflop/s)	125
maximal (Gflop/s)	8
Comm. bandwidth	
bidirectional point-to-point (MB/s)	40
bisectional (GB/s)	1.28
No. of processors	8-64
Memory/node (MB)	64-256

Recently, a second member of the Scalable Parallel family, the IBM SP2 has been introduced, which is based on IBM POWER2 technology.

Machine type: RISC-based distributed-memory multi-processor

Model: IBM 9076 SP2

Clock cycle (ns)	15
Theor. peak performance	
per proc. (Mflop/s)	266
maximal (Gflop/s)	34
Comm. bandwidth	
bidirectional point-to-point (MB/s)	40
bisectional (GB/s)	2.5
No. of processors	8-128
Max memory/node (MB)	64-2048

The programming model is SPMD (Single Program Multiple Data) with explicit message-passing.

Message passing is supported by MPL [145] (Message Passing Library) IBM's proprietary, and PVMe [146], an IBM's proprietary library enhancing the PVM performance on SP platforms, while ensuring compatibility with the public domain version PVM communication library.

A.2 The nCUBE architecture

The nCUBE systems are hypercubes based on custom designed 64-bit VLSI CMOS 1-processor nodes.

Machine type: Distributed-memory multi-processor

Model: nCUBE 2S

Connection structure: Hypercube

Clock cycle (ns)	40
Theor. peak performance	
per proc. (Mflop/s)	3
maximal (Gflop/s)	20
Comm. bandwidth (MB/s)	2.75
No. of processors	8-8192
Max memory/node (MB)	64

The VLSI chip, together with DRAM memory, constitutes what nCUBE calls a **computing element**. Each processing element is connected to the system network for message passing through **D**irect **M**emory **A**ccess (DMA) channels controlled by the processor. In this context, a channel is either a bidirectional data path between contiguous nodes or a node-to-I/O data path.

Channels operate at 22.2 MB/s nominal bandwidth. Each node owns 14 channels, 13 for interprocessor connections ($2^{13} = 8192$) and 1 for I/O.

A.3 The Intel Paragon XP

The Paragon is a commercial offspring of the experimental Touchstone Delta System. The Delta system used i860 processors as computational elements in its nodes, but, unlike its predecessor, the iPSC/860, the nodes were not arranged in hypercube topology but in a 2-D mesh.

The Paragon retains compatibility with the former iPSC/860 machine.

Machine type: RISC-based distributed-memory multi-processor

Model: XP/S

Connection structure: 2-D mesh (torus)

Clock cycle (ns)	20
Theor. peak performance	
per proc. (Mflop/s)	75
maximal (Gflop/s)	300
Comm. bandwidth (MB/s)	200
No. of processors	64-4000
Max memory/node (MB)	128

A.4 The CM-5 architecture

Machine type: Distributed-memory multi-vector-processor

Model: CM-5

Connection structure: fat-tree

Clock cycle (ns)	31.25
Theor. peak performance	
per proc. (Mflop/s)	128
maximal (Gflop/s)	2048
Comm. bandwidth (MB/s)	5-20
No. of processors	16-16384
Max memory/node (MB)	32

The CM-5 is a scalable distributed memory machine in which each node is a 32 MHz Sun SPARC processor with four optional vector units.

Two internal networks are available to support interprocessor communication: the Control Network and the Data Network. The Control Network is responsible for global, or collective, communications in which many processors handle a few data, such as global reductions and processor synchronization.

The Data Network supports point-to-point communication, and is implemented as a fat-tree.

A.5 The Cray T-3D architecture

The T3D is the first generation of distributed memory MIMD systems from Cray Research Inc. The T3D processing node is built around a dual 150

MHz Alpha processor. A 3D torus interconnection supports bidirectional traffic along the three directions x, y, z.

The T3D supports "virtual addressing", meaning that it can address global data residing outside local memory without generating interrupts to the remote processor owning the data to be moved in.

Machine type: RISC-based distributed-memory multiprocessor

Model: T3D

Connection structure: 3-D torus

Clock cycle (ns)	6.6
Theor. peak performance	
per proc. (Mflop/s)	150
maximal (Gflop/s)	307
Comm. bandwidth (MB/s)	300
No. of processors	32-2048
Max memory/node (MB)	64

A.6 The SGI Power Challenge architecture

The Challenge systems are shared-memory MIMD computers using 75 MHz MIPS R8000 processors. The data are transported from the main memory to the CPUs by a central bus 256 bits wide, with a 1.2 GB/s bandwidth.

One of the main appealing features of the SGI system is that parallelisation can be done automatically by the compiler (Fortran or C). Moreover, the Power Challenge architecture provides hardware support for fast synchronization operations to allow for efficient fine-grain parallel processing.

Machine type: Shared-memory multiprocessor

Model: Power Challenge

Clock cycle (ns)	13.3
Theor. peak performance	
per proc. (Mflop/s)	300
maximal (Gflop/s)	5.4
Memory bandwidth (GB/s)	1.2
No. of processors	2-18
Max memory/node (MB)	16

Bibliography

[1] L. Landau, E. Lifshitz, 'Fluid Mechanics', Pergamon Press, London, 1959

[2] L. Richardson, 'Weather prediction by numerical process', Cambridge University Press, Cambridge, 1922

[3] G. E. Karniadakis, S. Orszag, 'Nodes, modes and flow codes', Phys. Today, 46,3,1993

[4] P. Cvitanovich, 'Universality in chaos', World Scientific Singapore, 1984

[5] U. Frisch, S. Orszag, 'Turbulence: challenges for theory and experiment', Phys. Today, 46, 3, 1993

[6] A. Isihara, 'Statistical Physics', Academic Press, London, 1971

[7] A. Kolmogorov, Dokl. Acad. Nauk, SSSR 30, 299, 1941

[8] H. Tennekes, J. Lumley, 'A first course in Turbulence', MIT Press, Cambridge, 5th ed., 1978

[9] B. Launder, D. Spalding, 'Mathematical models of Turbulence', Academic Press, 1970

[10] R. Rogallo, P. Moin, Ann. Rev. Fluid Mech., 16, 1984

[11] U. Frisch, Proceedings Microscopic Simulation of Complex Hydrodynamic Phenomena, NATO ASI Series B, Physics, Vol. 292, M. Mareschal, B. Holian eds, 1992

[12] U. Frisch, 'Turbulence: the legacy of A.N. Kolmogorov', Cambridge Univ. Press, 1995

[13] E. Oran, J. Boris, 'Numerical simulation of reactive flows', Elsevier, New York, 1987

[14] F. Williams, 'Combustion Theory', Redwood City, Bejamin/Cummings, 1985

[15] R. Agarwal, 'Parallel computers and large problems in industry', Comp. Meth. in App. Science, Elsevier, 1, 1992

[16] A.J. van der Steen, 'Overview of recent supercomputers', Nationale Computer Faciliteiten Report, Academic Computing Centre Utrecht, Sept. 1994

[17] J. Dongarra and B. Tourencheau eds, Proceedings 2nd workshop on Environments and Tools for Parallel Scientific Computing, SIAM, Philadelphia, 1994

[18] J. Jimenez, P. Moin, 'The minimal flow unit in near wall turbulence', J. Fluid. Mech. 225, 240, 1991

[19] C. Cercignani, 'Theory and Applications of the Boltzmann Equation', Scottish Academic Press, 1975

[20] C. Cercignani, R. Illner, M. Pulvirenti, 'The mathematical theory of dilute gases', Springer Verlag Series in Applied Mathematics, vol. 106, 1994

[21] G. Batchelor, 'An Introduction to Fluid Dynamics', Cambridge Univ. Press 615, 1975

[22] R. Peyret, 'Computational methods for Fluid Flows', Springer Verlag Series in Computational Physics, 1986

[23] S. Patankar, 'Numerical heat transfer and fluid flow', Mc Graw Hill, 1980

[24] R. Issa, A. Gosman, A. Watkins, 'The computation of compressible and incompressible recirculating flows by a non-iterative implicit scheme' J. Comp. Phys. vol. 62, p. 66, 1986

[25] A. Amsden, P. O'Rourke, T. Butler, 'KIVA-II: A computer program for chemically reactive flows with sprays', Los Alamos National Laboratory report LA-11560-MS, 1989

[26] C. W. Hirt, A. A. Amsden, J.L. Cook, J. Comp. Phys. 14, 227, 1974

[27] C. Fletcher, 'Computational techniques for fluid dynamics I-II', Springer Series in Computational Physics, 1990

[28] J. Ferziger, M. Peric, 'Computational Methods for Fluid Dynamics' Springer Verlag Series in Applied Mathematics, 1995

[29] A. Chorin, 'Vorticity and Turbulence', App. Math. Sciences, vol. 103, Springer Verlag 1994

[30] D. Mansutti, G. Graziani, R. Piva, 'A discrete vector potential model for unsteady incompressible viscous flows', J. Comp. Phys., 92, 1991

[31] J. Boris, 'New directions in Computational Fluid Dynamics', Ann. Rev. Fluid Mech., 21, 695, 1987

[32] S. Patankar, 'Numerical heat transfer and fluid flow', Mc Graw Hill, 1980

[33] J. Ferziger, M. Peric, 'Computational Methods for Fluid Dynamics' Springer Verlag Series in Applied Mathematics, 1995

[34] C. Canuto, M. Hussaini, A. Quarteroni, T. Zang eds, 'Spectral methods in Fluid Dynamics', Springer Verlag, Berlin, 1988

[35] D. Potter, 'Computational Physics', Wiley London, 1973

[36] C. Birsdall, A. Langdon, 'Plasma Physics via Computer Simulation', Mc Graw Hill, 1985

[37] J.F. Thompson, Z.U.A. Warsi, C. Wayne Mastin, 'Numerical grid generation', Elsevier Science Publishing Co., New York 1985

[38] A. Jameson, 'The present status, Challenges and Future Developments in Computational Fluid Dynamics', AGARD Symposium, CP-578, 1996

[39] G. Strang, G. Fix, 'An analysis of the Finite Element Method', Prentice Hall, Englewood Cliffs, 1973

[40] O. Pironneau, 'Finite Elements Methods for Fluids', John Wiley, New York, 1989

[41] F. Brezzi, M. Fortin, 'Mixed and Hybrid Finite Element Methods', vol. 15, Springer Verlag Series in Computational Mathematics, 1991

[42] K. Morton, M. Baines eds, 'Numerical methods for Fluid Dynamics', Oxford, Clarendon, 1986

[43] D. Gottlieb and S. Orszag, 'Numerical Analysis of Spectral Methods: Theory and Applications', Philadelphia, SIAM, 1977

[44] S. Orszag, 'Comparison of pseudospectral and spectral approximation', Stud. App. Math. 51, 253, 1972

[45] P. Moin, J. Kim, 'Numerical investigation of turbulent channel flow', J. Fluid Mech., 118, 341, 1982

[46] A. Patera, 'A spectral element method for fluid dynamics', J. Comp. Phys., 54, 468, 1984

[47] P. Fischer, A. Patera, J. Comp. Phys., 92, 380, 1991

[48] G. Marchuk, 'Methods of Numerical Mathematics', Springer Verlag Series in Applied Mathematics, vol. 2, 1975

[49] I. Duff, A. Erisman, J. Reid, 'Direct Methods for Sparse Matrices', Monographs on Numerical Analysis, Oxford Science Publications, 1989

[50] G. Golub, C. Van Loan, 'Matrix Computations', Johns Hopkins Univ. Press, 2nd edition, 1987

[51] H. Hageman, D. Young, 'Applied Iterative Methods', Academic Press, New York, 1981

[52] M. Hestenens and E. Stiefel, 'Methods of Conjugate Gradient for Solving Linear Systems', J. Res. Nat. Bureau Standards, B. 49, 409, 1952

[53] G. Radicati, Y. Robert, S. Succi, 'Iterative Algorithms for the Solution of Nonsymmetric Systems in the Modeling of Weak Plasma Turbulence', J. Comp. Phys., 80, 489, 1989

[54] R. Issa, A. Gosman, A. Watkins, 'The computation of compressible and incompressible recirculating flows by a non-iterative implicit scheme' J. Comp. Phys. vol. 62, p. 66, 1986

[55] R. Hockney, J. Eastwood, 'Computer Simulations using Particles', Mc Graw Hill, New York, 1981

[56] B. Alder, T. Wainwright, 'Molecular Dynamics by Electronic Computers', Proc. Transport Processes in Statistical Mechanics, I. Progogine ed., Interscience, New York, 1958

[57] W. Hoover, 'Computational Statistical Mechanics', Elsevier, Amsterdam, 1991

[58] M. Allen, D. Tildesley, 'Computer Simulation of Liquids', Oxford Univ. Press, Oxford, 1987

[59] A. Rapaport, E. Clementi, Phys. Rev. Lett., 57, 695, 1987

[60] M. Mareschal, E. Kestemont, Nature, 323, 427, 1987

[61] D. Rapaport, Computer Phys. Comm., 62, 2 and 3, 198, 1991

[62] G. Bird, 'Molecular Gas Dynamics', Oxford Univ. Press, London, 1976

[63] G. Bird, 'The perception of numerical methods in Rarefied Gas Dynamics', Progr. in Astronautics and Aeronautics, 118, 1989

[64] U. Frisch, B. Hasslacher, Y. Pomeau, 'Lattice gas Automata for the Navier-Stokes Equations', Phys. Rev. Lett. 56, 1505, 1986

[65] D. Dab, A. Lawniczak, J.P. Boon, R. Kapral, Phys. Rev. Lett., 64(20) , 2462, 1989

[66] D. Rothman, S. Zaleski, 'Lattice Gas Models of phase separation: interfaces, phase transistions and multiphase flows', Rev. Mod. Phys., 66, 1417, 1994

[67] T. Toffoli and N. Margolus, 'Programmable matter: concepts and realization', Procs Lattice gas methods for PDE's: theory applications and hardware, Physica D, 47, 1-2, 263, 1991

[68] R. Brandt, 'A new way to see the way the wind blows', MIT Reporter, Technology Review Magazine, Feb/March 1993

[69] R. Benzi, S. Succi, M. Vergassola, 'The Lattice Boltzmann equation: theory and applications', Phys. Rep. vol. 222, 3, p. 145, 1992

[70] Y. Qian, S. Succi, S. Orszag 'A review of the Lattice Boltzmann Method', Ann. Rev. Comp. Phys., vol. 3, p. 195, 1995

[71] 'Lattice Gas Methods: Theory, Applications and Hardware', G. Doolen ed., MIT Press, Cambridge, 1991

[72] A. Chorin, 'Numerical study of slightly viscous flow', J. Fluid Mech., 57, 785, 1993

[73] C. Marchioro, M. Pulvirenti, 'Vortex methods in two-dimensional fluid mechanics', Springer Verlag, New York, 1984

[74] L. Greengard, 'The numerical solution of the N-body problem', Computers in Physics, March/April, 142, 1990

[75] L. Lucy, Astron. J., 82, 1013, 1977

[76] J. Monaghan, Ann. Rev. Astron. Astrophys., 30, 543, 1992

[77] W. Hoover, T.Pierce, C. Hoover, J. Shugart, C. Stein, A. Eduards, 'Molecular Dynamics, Smoothed-Particle Applied Mechanics and Irreversibility', Comp. Math. Appl., vol. 28, n. 10-12, p. 155, 1994

[78] W. Hoover, A. de Groot, C. Hoover, 'Massively parallel computer simulation of plane-strain elastic flow via non-equilibrium molecular dynamics and Lagrangian continuum mechanics', Computers in Physics, 6, 155, 1992

[79] L. Verlet, 'Computers Experiments on Classical Fluids, I) Thermodynamic Properties of Lennard-Jones Molecules', Phys. Rev. 159, 98, 1967

[80] C. Gear, 'Numerical Initial Value Problems in Ordinary Differential Equations', Prentice Hall, Engelwood Cliffs, NY, 1971

[81] R. Hockney, Math. Comp. Phys., 9, 136, 1970

[82] A. Leonard, 'Computing 3D incompressible flows with vortex elements', Ann Rev Fluid Mech, 17, 523, 1985

[83] H. Lorin, 'Introduction to computer architecture and organization', John Wiley and Sons, Wiley Interscience, 1989

[84] J. Hecht, 'Physical limits of computing', Computers in Physics, 34, July/August 1989

[85] P. Kogge, 'The architecture of pipelined computers', Mc Graw Hill, New York, 1981

[86] W. Gentzsch, 'Vectorization of computer programs with applications to computational fluid dynamics', Notes on Numerical Fluid Mechanics, Vieweg, vol. 8, 1984.

[87] R. Hockney, C. Jesshope, 'Parallel Computers', Adam Hilger, Bristol, 1981

[88] D. Bayley, 'RISC Microprocessors and Scientific Computing', Proceedings Supercomputing 93, 645, 1993

[89] R. Oehler, R. Groves, 'IBM RISC System/6000 Processor Architecture', IBM J. Res. and Dev., 34,1, 1990

[90] K. Dowd, 'High performance Computing, RISC architecture, optimization and benchmarks', O'Reilly and Associates Inc., 1993

[91] M. Flynn, 'Some Computer Organizations and their Effectiveness', IEEE Transactions on Computers, C-21, 948, 1984

[92] G. Fox, M. Johnson, C. Lyzenga, S. Otto, J. Salmon, D. Walker 'Solving Problems on Concurrent Processors', vol.1, Prentice Hall, Englewood Cliffs, NJ, 1988

[93] T. Feng, 'A survey of interconnection networks', IEEE Computer, 12-27, 1981

[94] V. Kumar, A. Grama, A. Gupta, G. Karypis, 'Introduction to Parallel Computing', The Benjamin/Cummings Publ. Co., 1994

[95] I. Foster, 'Designing and building parallel programs', Addison Wesley, 1995

[96] G. Almasi, A. Gottlieb, 'Highly Parallel Computing', The Benjamin/Cummings Publ. Co., 1989

[97] G. Fox, W. Furmanski, Proceedings Scientific Applications of Multiprocessors, C.A.R. Hoare Series, Prentice Hall International, Series in Computer Science, 1989

[98] K. Reinsch,W. Schmidt, A. Ecer, J. Hauser, J. Periaux eds, Parallel Computational Fluid Dynamics '91, North Holland, 1992

[99] R. Pelz, A. Ecer, J. Hauser eds, Parallel Computational Fluid Dynamics '92, North Holland, 1993

[100] R. Sicovec, D. Keyes, M. Lenze, L. Petzold, D. Reed, 'Parallel processing for scientific computing', SIAM Philadelphia, 1993

[101] W. Gropp, 'Parallel computing and domain decomposition', 5th Int. Symp. on Domain Decomposition methods for partial differential equations, chap. 29, SIAM 1992

[102] B. Boghosian, 'Computational physics on the Connection Machine', Computers in Physics, Jan/Feb 14, 1990

[103] R. Eigenmann, 'Parallel architectures and how to program them', Proceedings 16th Workshop Speedup, vol. 8/2, p.39, 1994

[104] K. Tsuboi, K. Miyakohsi, K. Kuwahara, 'An approach to incompressible flow problems with a massively parallel computer', Proceedings of Parallel CFD 91, North-Holland, 343, 1991

[105] W. Gropp, E. Smith, 'Computational fluid dynamics on parallel processors', Computers and Fluids, vol. 18, p. 289, 1990

[106] G. Fox, 'Hardware and software architectures for irregular problems', in "Unstructured scientific computation on scalable multiprocessors", J. Mehzotre, J. Saltz, R. Voigt eds, MIT press, Cambridge 1992

[107] 'Partitioning of unstructured problems for parallel processing', Comput. Sys. in Engineering, 2, 135, 1991

[108] C. Fahrat, 'On the mapping of massively parallel processors onto finite element graphs', Computers and Structures, 32(2), 347, 1989

[109] R. Pelz, 'The parallel Fourier pseudo-spectral method', J. Comp. Phys., 92, 296, 1992

[110] P. Fischer, A. Patera, 'Parallel simulation of viscous incompressible flows', Ann. Rev. Fluid Mech., 26, 483, 1994

[111] E. Bonomi, M. Tomassini, 'Massively data-parallel molecular dynamics', Int. J. Mod. Phys. C vol. 3, 4, 709, 1992

[112] D. Rapaport, Computer Phys. Comm., 62, 2 and 3, 198, 1991

[113] W. Camp, S. Plimpton, B. Hendrickson, R. Leland, 'Massively parallel methods for engineering and science problems', Comm. of the ACM, 37, 4, April 1994

[114] L. Dagum, 'Implementation of hypersonic rarefied flow particle simulation on the Connection Machine', Proceedings Supercomputing '89, p.42, ACM 08971

[115] T. Bartel, S. Plimpton, 'Direct Montecarlo Simulation of ionized flows on large MIMD parallel supercomputers', Progress in Aeronautics and Astronautics, AIAA Series, vol. 159, 1994

[116] G. Kohring, 'Vectorization and Parallelization of Lattice Gas Algorithms', Int. J. of Modern Physics C2, 755, 1991

[117] M. Creutz, C. Rebbi, Phys. Rev. Lett., 42, 1390, 1979

[118] J. Barnes, P. Hut, 'A hierarchical NlogN force-calculation algorithm', Nature, 324, 446, 1986

[119] J. Sethian, 'Computational fluid mechanics and massively parallel processors', Proceedings Supercomputing '93, p.74, 1993

[120] J. Salmon, M. Warren, G. Winckelmans, 'Fast parallel tree-codes for gravitational and fluid dynamical N-body problems', Int. J. of Supercomp. App., 8, 2, 129, 1994

[121] J. Ortega, R. Voigt, 'Solution of partial differential equations on vector and parallel computers', SIAM, Philadelphia, 1985

[122] D. Evans, R. Sojoodi-Haghighi, 'Parallel Iterative methods for solving linear systems', Int. J. Comput. Math., 11, 247, 1982

[123] L. Adams, J. Ortega, 'A multicolor SOR method for parallel computation', Proceedings Int. Conf. Parallel Processing, 53, 1982

[124] D. Reed, M. Patrick, 'A model of asynchronous iterative algorithms for solving linear systems', Proceedings Int. Conf. Parallel Processing, 402, 1984

[125] D. Bayley, E. Barszcz, L. Dagum, H. Simon, 'NAS Parallel benchmark results', IEEE J., 43, 1993 and RNR Technical Report 94-006, March 1994

[126] R. Williams, 'Performance of dynamic load balancing algorithms for unstructured mesh calculations', Concurrency, Practice and Experience, 3(5), 457, 1991

[127] J. Lewis and R. Van der Geijn, 'Distributed memory matrix-vector multiplication and conjugate-gradient algorithms', Proceedings Supercomputing '93, p. 484, 1993

[128] I. Duff, S. Johnson, 'Node orderings and concurrency in structurally symmetric sparse problems', Proceedings Parallel Supercomputing: Methods, Algorithms and Applications, G. Carey editor, 177, NY Wiley, 1989

[129] S. Breit, 'Implementing the incomplete LU factorization method on the BBN TC2000 parallel computer', Proceedings Parallel CFD 91

[130] R. Keunings, O. Zone, R. Aggarwal, 'Parallel algorithms in computational rheology', Proceedings Theoretical and Applied Rheology, P. Molderhauer ed., 274, Elsevier, Amsterdam, 1992

[131] W. Gropp, D. Keyes, 'Domain Decomposition Methods in Computational Fluid Dynamics', Int. J. Num. Meth. Fluids, 14, 140, 1992

[132] A. Quarteroni, 'Domain decomposition and parallel processing for the numerical solution of partial differential equations', Survey Math. Ind., 1, 75, 1991

[133] A. Quarteroni, J. Periaux, Y.A. Kuznetsov, O.B. Widlund, 'Domain decomposition methods in science and engineering', Contemporary Math. 157, SIAM, 1994

[134] F. Dellagiacoma, S. Paoletti, F. Poggi, M. Vitaletti, 'Multidomain computations of compressible flows in a parallel scheduling environment', Proceedings Parallel CFD 92, R. Pelz, A. Acer, J. Hauser eds, p. 111, 1993

[135] T. Michl, S. Maier, S. Wagner, M. Lenke, A. Bode, 'The parallel implementation of a 3D Navier-Stokes solver on Intel Multiprocessor systems', Proceedings 14th Workshop on Vector and Parallel Computing, SPEEDUP, vol.7 n. 2, p.19, 1993

[136] G. Punzo, F. Massaioli, S. Succi, 'High-resolution lattice Boltzmann computing on the IBM SP1 scalable parallel computer', Computers in Physics, vol. 8, n.6, 1994

[137] S. Succi, B. Ayati, A. Hosoi, 'A six-lecture primer on parallel computing', University of Chicago Comput. Science Dept. Report CS96-11, April 1996

[138] F. Massaioli, R. Benzi, S. Succi, R. Tripiccione, Eur. J. Mech., B Fluids, 14, n 1, 1995

[139] G. Bachler, M. Bernaschi, R. Greimel, F. Papetti, S. Succi, 'Benchmarking the FIRE code on the IBM SP1/2 scalable parallel platforms', Proceedings HPCN95, Springer Verlag Series in Computational Science, p. 64, 1995

[140] M. Bernaschi, R. Greimel, F. Papetti, H. Schiffermuller, S. Succi, 'Numerical Combustion on a Scalable Parallel Platform', Applications in Advanced Architecture Computers, SIAM News, vol. 29, p. 16, June 1996

[141] S. Chen, X. Shan, 'High-resolution turbulent simulations using the Connection Machine 2', Computers in Physics, vol. 6, n.6, 1992

[142] J. Pal Singh , C. Holt, J. Hennessy, A. Gupta 'A parallel adaptive fast multipole method', Proceedings Supercomputing '93, ACM 0-8186-4340, p. 54, 1993

[143] P. Bhatnagar, E. Gross, M. Krook, Phys. Rev., 94, 511, 1954

[144] L. Long, J. Myczkowski, 'Solving the Boltzmann equation at 61 Gflop/s on a 1024-node CM5', Proceedings Supercomputing '93, ACM 0-8186-4340, p.528, 1993

[145] IBM AIX Parallel Environment, Operation and Use SH26-7230, 1993

[146] IBM AIX PVMe, User's Guide and Reference SH23-0019, 1994

Subject Index

A

Algorithm, Reverse Cuthill-McKee, 137
Arbitrary Lagrangian Eulerian, 24
Architecture, Cray T-3D, 218
Architecture, IBM/SPI/SP2, 215
Architecture, n CUBE, 216
Architecture, SG1 Power Challenge, 219
Architecture, Superscalar RISC, 92
Architecture, Vector 90
Architectures, Parallel Computer , 93, 215
Architectures, Superscalar, 92

B

Basic equations of Fluid flow, 15-30
BBGKY Hierarchy, 16
Blocking, 181-182
Boltzmann equation, 17

C

Caching, 92
Chaining, 91
CM-2, Parallel Spectral Codes for Homogeneous Incompressible Turbulence on the, 205
CM-5 Architecture, 218
Coherent Structures, 5
Communication Computation Ratio, 110
Communication overheads, Coordination and, 109
Compressibility, 25
Computer networks, 98
Computers, Role of, 10
Computers, Superscalar, 92
Computers, Vector, 88
Conjugate Gradient, 62
Control mechanism, 94
Convection, 3
Coordination and Communication overheads, 109
Courant-Fredricks-Levy Conditions, 55
Cray T-3D Architecture, 218

D

Dependency Segregation, 183
Direct methods, 58, 178, 184
Direct Monte Carlo Simulation, 69, 83, 155
Discrete Space-Time Grid, 32
Dissipation, 3, 4
Distributed Memory M/MD, 125
Domain Decomposition, 124
Dynamic Load balancing, 108
Dynamic networks, 98, 100

E

Eddy viscosity. 6
Efficiency of Structured Explicit Schemes, 128
Energy Conservation, 19
Engine Combustion with Unstructured conjugate-gradient solver, 202

Enhanced transport, 6
Eulerian form, 21
Explicit / Implicit, 193

F

Far-Field Approximation, 160
Fast Multiple Methods, 162
Fast Summation Techniques, 209
FCHC Lattice, 75
FHP Lattice Gas, 73
Finite difference methods, 35
Finite Element Method, 43
Finite Elements, 44
Finite Volume Mesh, 38
Finite Volume method, 38
Fluid dynamics research in Science and engineering, 1-14
Fluid Flow Equations, Numerical Solution of, 31, 65-84
Fluid flow, Basic equations of, 15-30
Force Decomposition, 153
Fraction of parallel content, 105

G

Gauss-Seidel Scheme, 60
Geometry, role of, 7-8
Global Communication, 171
Global Matrix, 49
Global Methods, 34
Gradient Methods, 61, 175
Greedy Algorithm, 137
Grid methods vs. Particle methods, 32
Grid methods, 34

H

Halo cells, 126
Homogeneous Incompressible Turbulence on the CM-2, Parallel Spectral Codes for 205
Hydrodynamics, Smoothed Particle, 80, 84
Hypercubes, 101

I

IBM SP1, Lattice Boltzmann on the, 197
IBM/SPI/SP2 Architecture, 215
Incomplete Nested Dissection, 190
Incompressible Flow past Solid Bodies with Parallel Vortex Methods, 207
Intel Paragon XP, 217
Interprocessor Communication, 126, 199
Irrotational Flow, 29
Iterative methods, 59, 184
Iterative/Direct, 193

J

Jacobi Scheme, 60

L

Lagrangian Formulation, 23
Lattice Boltzmann on the IBM SP1, 197
Lattice Boltzmann, 76, 82, 157
Lattice Gas, 72, 82, 156
Lennard-Jones Potential, 68
Linear algebraic equations, 57
Load balance, 108
Load balancing, Dynamic, 108
Local Mesh Refinements, 43
Local Methods, 34
Locality, 36

Loval Matrix, 49
LU Decomposition, 179

M

M/MD, Distributed Memory, 125
Mass Conservation, 18
Matrix Bandwidth, 37, 59
Matrix Partitioning, 124
Maximally Discretized Molecular Dynamics, 75
Memory organization, 95
Mesh Networks, 99
Mesh Numbering Strategies, 49
Message Coalescence, 200
Molecular Dynamics, 66, 82, 148
Molecular Dynamics, Maximally Discretized, 75
Momentum Conservation, 19
Monte Carlo, Parallel Direct Simulation, 155
Multicolored Block-Iterative Methods, 172
Multicoloring, 167
Multiple Instruction Multiple Data, 95
Multiple Instruction Single Data, 95
Multipole Expansions, 159, 161

N

n CUBE Architecture, 216
N-Body Problems, Numerical, 158
Navier-Stokes equations, 18-21
Neumann computer, 85-86
Non-linear systems, 2
Non-linearity, 57
Non-linearity, Time-discretization and , 53
Numerical N-Body Problems, 158
Numerical Solution of Fluid Flow Equations, 31, 65-84

P

Parallel Computer Architectures, 93, 215
Parallel computing, 85-118
Parallel Direct Simulation Monte Carlo, 155
Parallel DST Scheme, 207
Parallel Explicit Grid Techniques, 121
Parallel Implicit Methods, 165-196
Parallel iterative schemes, 165
Parallel programming, 102
Parallel Rarefied Gas Dynamics, 209
Parallel Spectral Codes for Homogeneous Incompressible Turbulence on the CM-2, 205
Parallel Vortex Methods, Incompressible Flow past Solid Bodies with , 207
Parallelization of Matrix-Vector-Multiply, 138
Parallizing Backward/Forward Substitution, 182
Particle Methods, 65-79, 194
Particle methods, Grid methods vs, 32
Particle Methods, Time Machine with, 82
Perfect Fluid, 22
Picard's Scheme, 60
Pipelining, 88-89
Potential flows, 29
Pseudo Spectral Methods, 52

R

Rarefied Gas Dynamics, Parallel, 209
Reactive Flows, 9
Recursive Graph Bisection, 137
Recursive Spectral Bisection, 138
Redundancy, 116
Relaxation schemes, 60, 165
Reverse Cuthill-McKee Algorithm, 137
Reynolds Number, 3
RISC Architecture, Superscalar , 92
RISC processor, Superscalar , 93

S

Scalability, 103
Scalability, 11
Scattered Box Frontal Method, 189
Science and engineering, Fluid dynamics research in, 1-14
SG1 Power Challenge Architecture, 219
SIMD, 95, 96
SIMD, Distributed Memory , 128
Simulation Monte Carlo, Parallel Direct 155
Single Instruction, Multiple Data, 95
Single Instruction, Single Data, 95
SISD-95
Smoothed Particle Hydrodynamics, 80, 84
Spatial Decomposition, 149
Spectral Codes for Homogeneous Incompressible Turbulence on the CM-2, 205
Spectral Methods, 50, 141, 194
Static networks, 98, 99
Steepest Descent (SD), 62
Structured Explicit Schemes, Efficiency of , 128
Structured Explicit, 121
Structured/Unstructured, 193
Substructuring, 184
Superscalar architectures, 92
Superscalar computers, 92
Superscalar RISC Architecture, 92
Superscalar RISC processor, 93

T

Time Machine with Particle Methods, 82
Time-discretization and non-linearity, 53
Time-discretization and Non-linearity, 53
Topology, 97
Turbulence Modelling, 7
Turbulence on the CM-2, Parallel Spectral Codes for Homogeneous Incompressible, 205
Turbulence, 2-6
Turbulent flows, 8

U

Unstructured Explicit, 130
Unstructured Implicit Schemes, 188

V

Vector Architecture, 90
Vector Computers, 88
Vortex Methods, 78, 158
Vorticity Streamfunction Formulation, 26
Vorton Methods, 79/83